THE NEURO REVOLUTION

THE NEURO REVOLUTION

HOW BRAIN SCIENCE IS CHANGING OUR WORLD

ZACK LYNCH

WITH BYRON LAURSEN

St. Martin's Press New York

www.stmartins.com

Book design by Rich Arnold

Library of Congress Cataloging-in-Publication Data

Lynch, Zack.
The neuro revolution : how brain science is changing our world / Zack Lynch with Byron Laursen. — 1st ed.
p. cm.
Includes bibliographical references.
ISBN-13: 978-0-312-37862-2
ISBN-10: 0-312-37862-9
1. Neurosciences—Technological innovations. I. Laursen, Byron. II. Title.
RC337.L96 2009
612.8028—dc22

2008046321

First Edition: July 2009

10 9 8 7 6 5 4 3 2 1

For Casey and Kyle

CONTENTS

THE NEURO REVOLUTION

INTRODUCTION

INTO A NARROW TUNNEL

My first glimpse of humanity's future cost me plenty, as you'll see. But I can also tell you up front that it has turned out to be more than worth the price.

One morning in 1988 I stood motionless on the edge of a very small platform with the canopy of the north Queensland rain forest stretched to the horizon beneath me. I remember seeing moist warm tropical air dew up on my body. Then I dove into that Australian air. No shirt, no shoes, only pale blue shorts and a leather strap tightly fastening my ankles together. I plunged 140 feet into a hole where the trees had been cleared, swinging my arms wildly in circles, trying to maintain control. Seconds later, the tips of my then-long hair dipped gently into the watering hole that was my target. The bungee cord, attached to the leather around my ankles, had worked exactly as advertised. I was still in one piece. Now I could relax and enjoy the ride.

In an instant I shot back into the air. Once again I was looking at the rain forest canopy from above, though this time rising, feeling weightless, until gravity reasserted its force and I dropped to the water. My falls and ascensions gradually got shorter until finally I came to rest, dangling above the pool. What a rush! I had to do it again.

Within thirty minutes I was poised for a second jump. This time, to intensify the fun, I turned around, took a deep breath, bent my knees, swung my arms above my head, arched my back, and dove off backward. And that's what ripped my back to painful shreds. At the instant I hit bottom, it felt as if every molecule in my body teleported its weight to my lower back. An electric jolt zinged through my body, and with each of several successive bounces came another excruciating surge. The bungee had held, but critical parts of my spine had not.

Eight years of physical therapy and chiropractic care followed my bungee jump. I came to need an ever-escalating dose of pain medication just to get out of bed in the morning.

Eventually, a neurosurgeon ordered a sophisticated diagnostic scan of my lower back. I soon found myself being prepped at a University of California at San Francisco clinic to have my chronically agonized body checked out by imaging equipment that was, in 1996, state-of-the-art. And this is where I got an invaluable inkling of the future. It took quite some time and research before I began to grasp the importance of what I'd experienced, but that day at that clinic I took part in the early stages of a revolution that I believe will soon transform our future just as gigantically as certain other vital inventions—including the plow, the steam engine, and the microchip—did in our past.

The broad term is "neurotechnology," which means the tools we use to understand and influence our brain and nervous system. Beyond its medical applications, and most important for this story, is the fact that neurotechnology is driving advances in practically every field you can name, from finance and marketing to religion, warfare, and art. Because of its ever-widening impact, we are now entering something I've called the "neurosociety," an emerging reality where we will see massive changes in our personal, social, economic, and political lives, all of them driven by neurotechnology. As soon as I finish my bungee-cord/spinal-cord story, I'll start unveiling as complete a picture of the neurosociety as is now possible, drawing on the work and generously shared perspectives of some of the most brilliant and enthusiastic scientists, businessmen, and thinkers on the planet.

Stripped down to nothing but a loosely fitted hospital gown, two thick plugs stuck deep into my ears, I heard electric motors whirr as they sent me, headfirst and horizontal, into a cream-colored tube barely as wide as my shoulders. I was entering the clinic's magnetic resonance imaging (MRI) scanner.

The scanner bed stopped rolling. I was fully inside the tube. I heard a technician's voice through an overhead speaker, asking about my comfort and letting me know that the scanning was ready to commence. Some people say that being in an MRI machine sounds like a jackhammer. I'd describe it as more like the rapid thumping of room-sized pistons. At least it was quieter than the pounding drum 'n' bass electronic music I'd experienced a couple of years earlier at the Ministry of Sound, a dance club in London.

Thirty minutes later the scan was complete, the machine's computer had recorded and analyzed data about how those loud thumping noises had bounced back from the structures under my skin, and I was free to get dressed and hobble home. A few days afterward, at my doctor's office, I saw what the machine had noninvasively seen. A disk was ruptured in my lower back, pressing crucial nerves into a too-tight space. This was why I'd endured on-and-off agonies ever since the fateful jump. With evidence of progressive nerve damage in hand, the surgeon was ready to operate, confident that my injury was fixable. Through a small incision he dug out the offending disk, leaving my nerves free to return to their natural, unperturbed states. After relatively few months I was able to return—on a pain-free basis—to the slopes of the Sierras Nevadas with my snowboard, and to the South Pacific in my scuba-diving gear. Neurotechnology had played a major part in my liberation. I had to learn more.

The first whole-body MRI scan of a person was taken on August 28, 1980, in Scotland. The development of this imaging capability was soon recognized as the biggest improvement in medical diagnostics since the 1895 discovery of X-rays. Eventually, Paul Lauterbur and Sir Peter Mansfield won the Nobel Prize in Medicine for their discoveries that had made MRI possible.

In the early 1990s a new technique for the MRI emerged, called functional magnetic resonance imaging, or fMRI. They call it functional magnetic resonance imaging because, unlike the regular magnetic resonance imaging, which is used, for example, to see whether someone's knee cartilage needs repair, fMRI captures a sequence of activity while it is in progress. In terms of brain imaging, fMRI does this by taking a series of snapshots that reveal which specific regions of the brain are becoming active. An active region uses more oxygen and lights up more on an fMRI scan. That tells us which region starts working after a stimulus has occurred, like looking at a picture of cash or thinking about a certain topic, like love or beauty.

In just a few years, brain scanners have become far more powerful. Researchers from nearly every field have begun excitedly peering into human brains and learning stuff that has rocketed scientific progress and absolutely reshaped study and research priorities at major hospitals, clinics, universities, and corporate laboratories around the world.

Neurotechnology includes brain imaging systems as well as drugs and medical devices to treat the nearly 2 billion people worldwide who suffer from a neurological disease, psychiatric illness, or nervous system injury. But the medical side of these advances for conditions such as Alzheimer's disease, schizophrenia, depression, chronic pain and addiction is only one part, one of many startling and future-defining aspects of the neurotechnology revolution. Read Montague, an incredible, memorable, multifaceted scientist whom you'll meet more than once in upcoming pages, is just one of many brilliant men and women who foresee profound social consequences on the horizon: "Neuroscience writ large has the biggest peril stamp on it ever. If it [neurotechnology] really works, then it's like nuclear energy. And these technologies are going to mature faster than people imagine. I am stunned, myself, by how well we can eavesdrop and get practical measures of what's going on in people's heads. Today, we can tell whether or not you are thinking about yourself or somebody else. While this might sound crude, we previously had nothing like it. We wouldn't even have had this conversation ten years ago. So if it's accelerating, we're going to have a discussion of something equally surprising in just a few years."

Coincidentally, as I was reading a book by Montague and also thinking about the best way to launch you into the chapters ahead, I picked up the day's edition of the *New York Times* and read: "Attracted by the breadth of his interests, ranging from schizophrenia to music, Columbia University has appointed Oliver Sacks, the neurologist and writer, as its first Columbia artist, a newly created designation."[1]

Dr. Sacks has written ten books, including several popular works about brain science, each one brilliantly done. Probably best known are *The Man Who Mistook His Wife for a Hat* and *Awakenings*, which became a movie starring Robert De Niro and Robin Williams. He is also a frequent contributor to the *New Yorker*. His new appointment will allow him to roam freely across Columbia's departments. Lee C. Bollinger, president of Columbia, commented in the story that Dr. Sacks's appointment exemplified the university's effort to forge close ties between the brain sciences, the programs of the business and law schools, and the many schools of the humanities to ask and answer questions that are fundamental to our understanding of human existence.

Backing up Dr. Sacks's appointment was an additional $20 million to expand the study of neuroscience and make sure it includes this interdisciplinary approach. This money infusion came on the heels of a $200 million donation the previous year for building a new center to house the university's Mind, Brain and Behavior Initiative.

Clearly Columbia sees the coming wave of change that will be brought forth by new brain technologies. And it is most definitely not the only player in today's academic, cultural, political, and business arenas that has made the connection.

"Neuroscience could have an impact on the legal system that is as dramatic as DNA testing," proclaimed MacArthur Foundation president Jonathan Fanton, adding, "Neuroscientists need to understand law, and lawyers need to understand neuroscience." Putting its money where it counts, the foundation funded a $10 million multiuniversity effort in late 2007 to understand how neurotechnology is impacting the world's legal systems. "Neuroscientific evidence has already been used to persuade jurors in sentencing decisions, and courts have admitted brain-imaging evidence during criminal trials to support pleas of insanity," explains Michael Gazzaniga of the University of California, Santa Barbara, codirector of the project.

It isn't just universities that are seeing millions of dollars flow toward neurotechnology—although MIT did recently receive a $350 million gift to build the McGovern Institute for Brain Research—but also private-sector companies searching for ways to bring applications of the technology into our daily lives. Private venture capital investment in neurotechnology start-ups has skyrocketed 300 percent over the past decade, while at the same time the annual budget for the National Institutes of Health that focus on brain and nervous system diseases more than doubled, reaching nearly $7 billion. The bulk of this research funding is focused on understanding diseases and developing more-targeted treatments, but the knowledge being created from all of this work, as you will see in the coming chapters, is likely to accelerate a whole range of human endeavors, from trading stocks to appreciating art.

It has been eight years since I first sat down to produce this book. A major reason why it took so long is that keeping up with the growing flood of neuroscience literature is like trying to drink from a fire hose. As the noted neurobiologist Steven Rose said, "The worldwide effort being poured into the neurosciences is producing an indigestible mass of facts at all levels."

As I've learned more about the science and technology, I've developed a more scrupulous eye. It's a fine balance between guesstimating how technologies might advance and understanding the cultural reasons why they will or won't, a balancing act I have worked hard at nailing down so you can enjoy this journey with the confidence that you are being steered clear of most of the wreckage, and seeing the true must-see developments and possibilities.

Just as I stood on that platform and looked across the top of a vast rain forest nearly twenty years ago, this short introduction has put you on a platform that will let you look into an absolutely amazing, unbelievable, and yet thoroughly probable future. From the story of my day in the MRI lab you have already learned all the science necessary to begin understanding why so many brilliant minds are so highly enthusiastic about positive changes likely to come our way, and somewhere between cautious and fear-struck about the darker potentials that we will have to avoid. This coming transformation is a promise-versus-peril scenario. That's just one of many reasons I'm driven by the idea of getting more people to understand the size and scope of the changes coming our way. They will be huge, and will turbocharge development of the newer changes coming in their wake.

Managing this step in our own evolution intelligently and benevolently is going to be one of the most important things the human race will ever do.

ONE

TIME'S TELESCOPE

As I rolled into that narrow MRI tunnel in San Francisco, my body tightly surrounded by what was then the very newest scanning technology, I was like a caterpillar entering a chrysalis. My hope was to emerge eventually into a better existence, my chronic pain erased. That hope was fulfilled, something I'm still thankful for every day. But an even greater transformation was launched for me that day. I gradually came to see possibilities far greater than the positive impact of this superior imaging technology on medicine and surgery.

It was about recognizing a gargantuan historic inevitability: Vast changes are gathering from this new technology, propelling humanity toward a radical reshaping of our lives, families, societies, cultures, governments, economies, art, leisure, religion—absolutely everything that's pivotal to humankind's existence.

This gigantic wave of transformation will reach every corner of the planet. It will create a metamorphosis as complete as the changing of a larval worm into a butterfly.

If enough of us realize what is coming, and if we can infuse this emerging wave with practical and benevolent intelligence, continuously keeping our aspirations aimed high, it will let us create a future of greatly enhanced, better-balanced, and more satisfying individual lives within a vastly transformed society that we will build through an unimaginably powerful capability we've never had before: increasingly precise control over the most complex entity in the universe, the single most important determinant of the quality of the lives we lead—our human minds.

Scientists are now building a phenomenal body of knowledge, at an explosive rate, about how our brains respond the way they do, and why, and how we might leverage this accumulating knowledge into innovations that will impact every part of our lives. Better understanding our brains will lead to more solid and reliable decisions, as individuals and as nations, creating more lasting happiness. We will tap potentials humanity has dreamed of and reached for across aeons—to live comfortably, harmoniously, and prosperously with our physical environment, each other, and our own emotions. Literally knowing our own minds will create new ways of learning, working, distributing wealth, experiencing cultures, and being creative. We will be able to ease chronic pain on every level, from physical to spiritual. Life in this now-emerging neurosociety will be as advanced from current existence as the Renaissance was from the Stone Age. We will see enormous modification of our personal relationships, the bases of political power, expressions of art, religious experiences, modes of learning, physical and mental health, and business competitiveness.

Profound questions will arise all along the way, and major controversies will mount, as these personal and social transformations challenge deeply held beliefs about what it means to be human.

You may already have learned some aspects of neurotechnology and neuroscience from recent magazine and newspaper articles that have described how researchers now can see the workings of the human brain in real time. Most of these early stories are centered on medical possibilities. Medicine is both vital and fascinating, but it's really just one facet of the world's approaching transformation. Neuroscience now drives many fields of study. University department walls are shape-shifting as many brand-new combinations are being created outright, at a rapid pace. Neurotheology, neurolaw, neuromarketing, neuroesthetics, and neurofinance are among the examples. These developments are morphing so quickly that even the most brilliant scientific minds at work today are often just vaguely aware of the neuroscience-propelled changes revolutionizing areas beyond their own fields of expertise.

The popular press reports on neuroscience that you may already have seen are your proof that the neurosociety has already taken baby steps. In the pages to come you'll find proof that the time is coming remarkably near for neuroscience to begin taking long, broad strides, and to become as unmistakable on our horizon of time as history's other great transformations did in their time.

I've spent the last several years positioned in full view of the coming attractions now being projected by some of the world's most brilliant minds. My job is to relentlessly track neuroscientific projects under way within the many businesses, universities, and independent labs now pursuing breakthroughs. My mission is to synthesize all the information that's flowing, shaping it into the best-informed ideas possible about where the rubber is going to meet the road.

The forthcoming chapters will reveal vital pieces of what I have seen from this unique vantage point, and your own visions of our possible tomorrows will take shape.

Change brings tumult, and inevitably it also inspires fear. The challenges ahead are tremendous. Deep social and cultural conflicts will arise, terrifying consequences may erupt, even as incredible benefits come about.

During the onrush of this neuroscience wave, it may seem at times—just as it often does today—that we are headed for a catastrophic future. Why shouldn't reasonable people be afraid of the future? Hellhounds are on our trail right now. Our news is a parade of horrific imagery: terrorists in continual and bloody resurgence; global climatic shifts; food shortages across the globe; the evaporation of the middle class; startling increases in suicide; radically volatile energy prices; human and financial assets hemorrhaging in wars being waged across the planet; currencies tanking; babies born daily into desperate poverty; visions taking shape of an unstable multipolar world. This hydra-headed malaise makes today's widespread prophecies of massive die-offs and possible extinction feel all too believable.

But thanks to the curiosity and drive of our ancestors, and to billions of us who will be working together in the near future within the neurosociety, we will be able to build a bridge wide enough for all of us to survive. We will also have the means to go higher than just survival. Depending on how we deal with the tumult, we will enter a flourishing age, riding a wave characterized by nearly unlimited access to the aspects of our humanity Abraham Lincoln summed up as "the better angels of our nature."

Since the dawn of civilization, humanity has undergone three societal revolutions. Each was driven forward by newly invented tools. Each of these technological leaps let people control the world around them to a far greater degree than was previously imaginable. Those surges of expanded control created three new epochs for humankind.

You are about to begin understanding the fourth.

Nearly ten thousand years ago, agricultural society came into being. Plows pulled by oxen replaced human muscles as the primary energy source for food production. Our ancestors were no longer forced to continually hunt, gather, and migrate. They began to gather surpluses. Sparse settlements grew into cities and city-states of hundreds of thousands of people. Specialized occupations were born, and the complexity of human life increased enormously.

Less than two and a half centuries ago, steam-powered engines became a reality, ushering in industrial society. Our control over energy creation, goods production, and resource distribution multiplied many fold. Distance became easier to conquer. New markets opened up around the world. The interconnectedness of human life was again magnified to a far higher degree.

In our time, the microchip gave birth to today's information society. We can tap into instantaneous global knowledge exchange. This accelerated communication and expanded access has created, in turn, vast new efficiencies across every existing industry, and the birth of industries and occupations that never existed before. The complexity and interconnectedness of all our lives has reached a staggering new level in a very short time.

These new technologies not only brought us new industries; they also reshaped business competition, personal communication, artistic expression, and warfare, bringing on such wide-ranging transformations that the lives of future generations were completely and forever changed.

Today we sit on the cusp of another overwhelming societal transformation, beginning to feel the liftoff of a wave potentially more dramatic than any of the three that came before. It is the emerging neurosociety. Early evidence of this wave will meet your eyes in the pages ahead. You will gradually realize that this coming wave will give us undreamed-of control of two vast spheres of life: both the *world around us* and the *universe within us.*

The forces driving the neurosociety's emergence are clear. Its arrival is both inevitable and already in progress. Even those who are now positioned closest to the unfolding wave cannot fully imagine the range and scope of impact on its way. It will be nothing less than the birth a new civilization.

Here is what I mean by inevitable: Global population has soared more than twentyfold over the past two hundred years, reaching over 6.6 billion. During the same two centuries, average life spans more than doubled, vaulting to more than seventy years. Current population projections say that the United States in 2040 will have 54 million people aged eighty-five and older, up from 4.2 million today. Today, those over eighty-five represent only 2 percent of the population. By 2040 they may represent almost 20 percent.

A population that is significantly older and massively larger, coupled with the recently created extensive global connectedness, has already created opportunities along with brand-new problems for modern humans. At the same time, it has intensified many of the old ones. We navigate our ever-changing lives with brains that have evolved very little since the Paleolithic Age. The problem-solving machinery in our heads is astonishingly complex, yet overwhelmed and overstimulated on a daily basis. It can turn quickly and insidiously, without our realizing it, into problem-causing machinery. We are constantly blasted with images of unattainable lifestyles, creating daily identity crises as we search for meaning in a world of continually shifting truths. Many of us are appalled by today's uneven distribution of wealth and power. Others are well supplied with both wealth and power yet are disillusioned, not able to feel the happiness such assets were supposed to provide. On every continent, in every culture, we see uncertainty, depression, anger, and resentment surfacing on a vast scale.

However, after spending thousands of years improving our control over the physical environment, we are about to receive new tools that will improve our control over the mental environment. These tools are a logical next step for helping conquer the stresses arising from living in our highly connected, urbanized information society.

Building on advances in brain science, neurotechnology (the set of tools for understanding and influencing the human brain) will allow us to experience life in ways never attainable before. Neurotechnology will enable people to consciously improve their emotional stability, enhance their cognitive clarity, and extend their most satisfying sensory experiences.

The Neuro Revolution will bring much more than fantastic new tools to enable individuals to experience a life less constrained by their evolutionarily influenced brain chemistry. It will deliver the capacity to reshape the very fabric, the innermost essential workings, of every industry, organization, and political system.

Let me share with you a vision of what is to come.

Using the last 250 years of history as our guide, we can use Time's Telescope to project forward over the next half century and see how businesses, governments, and personal relationships will shape and be shaped by humanity's fourth epochal transformation.

I developed Time's Telescope as a conceptual framework for looking at humanity's developments that is more refined than broadly descriptive epochs. It organizes recent human history into a succession of technological waves that build on the ones that came before them. I'll explain the model a little more thoroughly below, but I believe it gives us a reliable way to venture into long-term social forecasting. By forecasting I mean seeing what are the compelling forces at work, and what patterns they will follow as our future unfolds. Forecasting does not mean making highly specific predictions of finite events. It means mapping the direction and force of the coming wave in light of patterns that have occurred during previous periods of change.

There are good reasons for not making highly specific statements about what is to come. I believe in learning from history. History has shown that even the best-informed people are often terrible at predicting the future.

For example, in 1895 the eminent Irish physicist Lord Kelvin declared that heavier-than-air flying machines were "scientifically impossible." Five years later, the Wright brothers launched their test flights over the windy beaches of Kitty Hawk. Thomas Edison declared in 1880 that the phonograph he had invented had "no commercial value." I recently watched Herbie Hancock receive his Grammy for album of the year with *River: The Joni Letters*, so I'd have to join the millions who respectfully disagree with the Wizard of Menlo Park (the nickname Edison won for his frequent displays of genius). In 1955, a prominent home appliance manufacturer's CEO predicted that nuclear-powered vacuum cleaners would become a reality by 1965. Clearly, we are all happy that the atomic housecleaning revolution never took hold. In 1962, a Decca Recording Company executive turned down a young hopeful band by saying, "We don't like their sound. Groups of guitars are on the way out." So four rather scruffy guys from Liverpool, collectively known as the Beatles, had to keep knocking on doors. Then they blew the roof off of pop culture and the music business worldwide. Early in the 1970s, experts told us that Japan's electronics industry would soon reach a dead end because the market for stereos and transistor radios was nearly saturated. Examples like this are littered throughout the history of the future-predicting business. Given the now-obvious bonehead errors of these previous projections, where should we look for a fairly clear understanding of where society is headed in the future?

My training is to look at history. Of course, most of us know Napoleon's famous comment that history is just "the lies agreed upon by the winners," and we also realize that there are ways a person can divide up history to support almost any argument. Time's Telescope is just one of many ways of viewing history, but the perspective it provides is compelling.

Looking back over the 250 years that have passed since the first spark of the Industrial Revolution, we can see that newly developed technologies have provided the cutting edge of societal transformation in a relatively consistent pattern of sixty-year waves of economic and political change. Each time, with each new wave, a new set of technologies has emerged to solve problems we previously believed were insurmountable.

Each of these five waves became driven by the development and widespread use of a few new low-cost products that allowed entirely new industries to be born and at the same time transformed old industries, creating new forms of social organization along the way.

When we trace the role of these critical technologies across the span of each wave, we can see a predictable pattern of change. These historical patterns have been extensively researched by great thinkers such as Nikolai Kondratieff, Brian Arthur, Christopher Freeman, and Carlota Perez. I believe the Neuro Revolution will evolve along the lines that these proven patterns illuminate. Here is why Time's Telescope gives us a solid foundation for the vital yet always risky enterprise of looking to the future.

The first wave, water mechanization, took place between 1770 and 1830. It gave us a huge jump in productivity and power by replacing handcrafted production with water-powered machinery. This first wave brought inexpensive cotton clothing and food to the masses.

The second wave, steam mechanization, began around 1820 and continued to about 1880. The machinery developed by the end of the first wave made inexpensive iron possible, leading to the massive build-out of iron railroads. Railroads accelerated our ability to get goods and services to distant markets.

The third wave, electrification, began in 1870 and continued to 1930 and made the production of steel far less expensive. Access to this superior metal transformed the railroad systems again, and also made the modern city possible. Steel, combined with the then-new electricity infrastructure, made skyscrapers, electric elevators, lightbulbs, telephones, and subways possible.

The fourth wave, motorization, arrived in 1910 and continued up to about 1970. Cheap oil ushered in mass assembly and the motorization of the industrial economy. Inexpensive transportation of goods and services was suddenly available to the masses. Cars, sometimes decried in their first years on the scene as "rolling bedrooms," changed nearly all aspects of our economic and social lives, leading to the build-out in the 1950s and beyond of extensive interstate highways and burgeoning suburbs.

The fifth wave, the information wave, began emerging around 1960. At first computers were limited to a small set of users, mainly universities, corporations, government, and the military. But gradually computers became smaller, more efficient, and more affordable. Individuals figured out some good reasons to bring them into their homes. The more they did so, the more momentum the information wave accumulated. Entrepreneurs and inventors pressed ahead to find the "killer applications," the products and services that would capitalize on the tremendous potential of the wave.

Futurists were already foreseeing and describing what they called the "information society" in the 1950s, but it took the actual economic expansion and social change that high technology delivered throughout the 1980s and 1990s to truly make it happen. Our lifetimes have been a window on tremendous geopolitical, economic, and social change. In less than fifty years information technologies have reshaped our world, making possible such radical developments as the fall of the Soviet Union, the rise of the Asian economies, the advent of real-time global capital networks, and the emergence of human-based asymmetrical warfare, wherein cell phones, garage door openers, and cheap riggings of ammunition—in the hands of people conspicuously willing to die—

have played hell with the most expensively equipped, technologically advanced military force in world history.

Each wave consists of four roughly fifteen-year periods, each of which begins with a technological irruption phase characterized by explosive innovation where whole new types of technologies emerge. This is quickly followed by a period of financial frenzy over the profit-making potential of these new technologies and the many industries they will spawn and reshape. A cursory look at previous waves shows that financial bubbles are common during this phase. Next, during the build-out phase, products expand to their market potential while profit margins decline substantially, leading to mass commoditization. Ultimately, growth slows, which leads to the irruption phase of the next set of world-changing technologies as investors seek new high-value-creating, high-profit-margin sectors to place their bets. We are currently nearing the end of the build-out phase of the information wave. It's far from being a spent force, but computerization has become so cheap and penetrated so deeply into our lives that it is becoming, in a sense, invisible.

These first five waves arrived in sixty-year cycles with decade-long overlaps at the beginning and end of each wave. As each advance rose, it built bigger and spread wider. It created fertile conditions that eventually germinated another advanced wave.

In the chapters ahead are some exquisite first signs of the sixth wave that is currently emerging. As a preview, I'd like to jump you ahead in time and share a few snapshots of the future in progress.

Advancing neuroscience has already begun to shake up our legal systems. The accuracy of brain scanning technologies for truth detection will soon eclipse existing polygraph "lie detector" tests, achieving accuracy rates of 90 or even 95 percent, good enough evidence to present in the Supreme Court. Mind reading will remain impossible, but systems that leverage brain imaging and verifiable emotion-sensing technologies to detect deception will be used during depositions, and will tilt the scale toward truth and justice. Unfortunately, in countries without protections of individual liberties, there is no telling how these technologies will be warped to solidify a monopoly on power.

Beyond these direct impacts, neuroscience may help us have a smoother path in the future, and truer justice, by enabling us to address crime's root causes. Neuroscience will help us find core truths about vitally important questions such as: Why does someone turn out to be a violent criminal? Is there a biological basis? The answer is yes, which leads me to believe that people will be sentenced with mind-altering drugs as an alternative to prison.

In previous waves, finance was one of the first industries to adapt to the new possibilities. The same thing will happen in the neurotechnology wave. Think of the billions of dollars that are on the line each moment in our tightly linked worldwide markets. The economic drivers for improved trading performance and greater accuracy in every financial decision are tremendous. Early neuroscientific findings have already shattered the economic dogma that people are rational economic actors. For example, recent research shows that we almost always overestimate the happiness that an event, like a purchase, will bring. We might believe a new BMW will make life much better, but no matter how great the car is, the reality of having it will be less exciting than we anticipated, and the excitement will fade quicker than we imagined. There will be multiple ways that conventional economic theory will be transformed by brain science. Beyond that, new tools will emerge to help financial professionals excel. Neurotech-enabled traders will have at least two new technologies at their disposal. One will be real-time brain scanning and neurofeedback software solutions that correlate previous brain states and trading successes to give traders a predictive capacity, based upon their continuously shifting neurobiology, of their potential in any given moment for scoring a new success. Another set of tools they're likely to use will be side-effect-free emotional stabilizers to maintain a calm state while executing high-stress, complex financial transactions.

Pushing further into the future, beyond what many today would consider to be reasonable, I expect the emergence of brain-computer interface systems that will expand an individual's capacity to parse data streams and accelerate profitable decision making. These new technologies will create a new playing field for those who have access to them. Authentic breakthroughs like these will alter cost structures and transform productivity across the global financial industry as well as many other competitive knowledge-based industries. Neurotechnology will radically transform the prevailing view of managerial common sense for how to achieve highest productivity and profitability.

Artistic expression and the entertainment industries will be altered just as amazingly during the Neuro Revolution. Electricity gave rise to the cinema, and information technology created video games. Neurotechnology will engender new forms of artistic development and appreciation. For example, virtual reality experiences, which are still in their infancy, will flourish to an incredible degree. They will include not only visionary landscapes and sound tracks to tantalize the senses, but emotion-sensing technologies that will adapt the entertainment experience to match the desires of the person having them. In another form of neurotech-driven convergence art, deeper knowledge about why we find certain depictions scary or funny will lead to the development of systems that are worn throughout a performance or VR game. These systems will magnify, through noninvasive magnetic stimulation across the scalp, specific emotional states. It will be fascinating to see how creative humans across the globe leverage these new emotion-provoking tools to fulfill our desires for new experiences.

Brain science is also elucidating the relationships between religion and the human mind. Some neurotheologists expect eventually to prove scientifically the existence of God. Others expect that they will give atheism full scientific legitimacy. Even if that monumental question is never answered, though, it is reasonable to assume that uncovering the neurobiological underpinnings of spiritual belief and experiences will give us provocative new answers and insights into realms that have always seemed beyond humanity's understanding. This new knowledge will shake the bedrock upon which many of today's world religions are built. It is already known that surgical implants for epilepsy, electrodes placed in a deep brain region called the angular gyrus, can spark out-of-body experiences. The Neuro Revolution will also produce noninvasive technologies that will be able to stimulate spiritual experiences at a distance, giving new meaning to "inspiring sermon." Advancing neurotheology may eventually help people who haven't devoted years of their life to meditation and prayer to achieve mystical and tranquil states. Beyond the mystical experiences themselves, there would be significant and possibly lasting aftereffects such as release from depression, better immune function, increased interconnectedness, and a more positive outlook on life.

Moving from the devotional to the destructive, the development of sophisticated neuroweapons is going to create a perpetual state of tension between promise and peril. In the development of neurowarfare, we will experience vast amounts of worry, debate, and conjecture over what the ultimate effects will be. Emotional detection systems will pervade public areas as global surveillance networks seek out terrorists and criminals. Enhancement of strength, stamina, and cognition for "warrior-athletes" of the future will represent the next form of combat readiness. These future warriors will be screened for performance potential and improved with next-generation enhancers that will make steroids and other controversial enhancers of today seem like St. Joseph's Aspirin for Children. And they will be armed, for example, with technologies to mute a person's memories.

New neuroweapon systems will be created quickly when neurotechnology becomes widespread as global demand for sophisticated neurotech-enabled entertainment systems and financial trading platforms drives rapid neurotech development.

Of course, some of these developments may not happen exactly as I describe. At this stage of the Neuro Revolution's emergence, it is not crucial that the exact details are predicted correctly, but rather that we are right about the expansive breadth of change coming into our lives within each of these vital areas.

As a launching point, an excellent first place for seeing how brain science is already reshaping our world is in the realm of free will and the criminal justice system. Did your brain make you do it?

TWO

THE WITNESS ON YOUR SHOULDERS

The first reward of justice is the consciousness that we are acting justly.

—Jean-Jacques Rousseau

The Devil made me do it the first time; the second time I done it on my own.

—Billy Joe Shaver

The city of Las Vegas has spent millions in the persistent promotion of its slogan "Whatever happens in Vegas stays in Vegas." The idea is to get you believing that a visit there gives you a license to act as wild as you might please, then head back home with absolutely no one getting wise to your escapades. It's a slogan that, as the comedian Bill Maher commented, comes about as close as the city can get to printing "Enjoy your hooker" on its official stationery. And in fact, while I'm writing this the current mayor of Las Vegas has been very vocal about his enthusiastic hopes for exceptionally elegant brothels to become another attraction anchoring the local economy.

But the secrecy his city advertises is absolutely bogus, an unfulfillable promise. Whatever happens in Las Vegas, or anyplace else, rides home with whoever took part in it—etched into several different areas of brain tissue. Memories of guilty acts come in more readily than music downloads on a computer drive, and are just about impossible to erase.

Detecting when people are telling the truth is a superheated issue. It impacts our legal system, and our society as a whole. Corporations and governments are already spending millions to find out whether neuroscience will radically improve our ability to tell reality from falsehood, terrorist from bystander, crook from innocent citizen, and how soon we will be able to make that magic happen. Whatever happens in Las Vegas is only a tiny slice, a few pixels in a worldwide picture. This expanding ability that we've recently gotten, to view the brain doing all its many data-processing jobs, is edging us closer to a day when deceit in a court of law, or anyplace else where we vitally need the truth to come out, may be impossible.

Neurolaw is an emerging field of study that seeks to explore the effects of discoveries in neuroscience on law and legal standards. There is so much activity and excitement right now about what neurolaw might do for truth detection, it would be easy to overlook the fact that neuroscience is also going to massively impact our legal system in several other ways. Truth detection is only one of them. Powerful research is in the works in many other areas of neurolaw, and they could bring about equally dramatic shifts within deeply stirring and controversial issues.

But truth detection is a perfect starting point for this chapter, for several reasons.

First, our courts are increasingly willing to weigh other neuroscience-based evidence in some of the most serious cases. At this writing, experts estimate that there are over nine hundred active cases in the United States in which neuroscience is an issue. Neuroimaging has been given as valid evidence in personal-injury suits, proving various kinds of harm and causation, including medical malpractice and toxic exposure. It has been used to terminate contracts when evidence indicated one of the parties had insufficient mental capacity to form a contract. In Illinois, a federal district court recently allowed brain-scanning evidence presented by the state to show a connection between kids playing violent video games and being overaggressive in real-life behaviors.

Second, the CIA and other intelligence agencies have been throwing millions at neuroscience researchers for a few years now, in hopes that they'll soon develop and deliver extremely advanced tools for protecting national security.

Homeland Security very recently began field tests of Malintent, a portable system that rapidly scans people, such as passengers headed for a plane, as they pass by an array of sensors meant to discover any mind-held intention to cause harm. The sensors can detect extremely fine nuances in involuntary movements of facial muscles, revealing attempts to hide thoughts. By 2010 these systems are also expected to be able to detect stress pheromones.

Third, science-minded entrepreneurs are building and launching business models based on emerging truth-detection technology, sensing billions they might make in the private sector. After all, experts estimate that private businesses pay for about four hundred thousand lie detector tests yearly, despite the fact that the technology is barely more reliable than flipping a coin. The government itself does an estimated forty thousand lie detector tests per year. So some hard-charging businesspeople are betting that there would probably be a lot more tests given, at expensive rates, for tests that were really effective and reliable. They don't even have to be perfect: The Supreme Court is willing to accept a 95 percent rate of accuracy.

Truth detection by neuroscientific means is a new phenomenon, and it has shown astounding accuracy. Researchers have demonstrated that remarkable potential exists for learning how to access our mental downloads and make our brains spill certain types of secrets. They do this, currently, by watching to see whether memory-related areas of the brain light up under very specific kinds of conditions—such as seeing evidence that would only resonate in the mind of someone who knows intimate details of a given crime.

There are researchers who say they have gotten results as good as 95 percent. Some expect to reach 100 percent within five years. If that happens, assessing the truth of a suspected criminal's alibi will become a quick, low-cost, nearly foolproof process. The court system will become far less congested. The wheels of justice won't always have to grind slowly. Sometimes they might seem to spin as fast as an automated pepper mill.

Or at least this is the scenario many researchers share, a dream that's inspired by real and richly significant progress. While they are moving determinedly along this path, one important truth of the matter is that a lot more progress has to happen before they can turn this scenario into fact. Another truth is that the pressure felt, the intensity of the wishing that propels this dream, runs so strong that we have to be extremely careful. The standard of proof for reliable truth detection has to be held very high, and we've historically shown that we will often settle for low standards.

There's a story told about land speculators approaching President Abraham Lincoln. They wanted to know which of the territories out west would be the next one to become a state. Land in a newly admitted state would jump in value. In return for his vital inside information, they offered a generous bribe. Lincoln said no. They were prepared for that response. They quickly added thousands of dollars more. He turned down that second offer, so they went astronomically higher. Lincoln then called a guard who stood nearby and said, "Please show these men out. They are getting close to my price."

The point of the story is this: If Lincoln was capable of feeling temptation's pull, so is everyone else in a human skin. Advancing neuroscience presents radically compelling possibilities to both our system of jurisprudence and our system of capitalism at the same time. Many among us have price points much lower than Abraham Lincoln's. When in its progress should we begin to trust neuroscientific lie detection?

More to the point, who are the researchers, entrepreneurs, and government officials we can rely on to share the truth about their accuracy in detecting truth?

Deception has always been a tremendously useful social skill. We all learn it in childhood, and use it often for the rest of our lives. Humans have been practicing deception since we lived in caves, and we still sometimes use it in hopes of owning bigger and better caves. So we're looking at a phenomenon that runs deep.

Additionally, even if neuroscience delivers testing with a 95 percent chance of being right, that bargain would still include a 5 percent chance of being wrong—a number that will loom extremely large to anyone who is falsely accused.

The fact that neuroscientific brain scanning isn't yet effective enough for routine use may be a good thing. Society needs some time to figure out what this new technology means, can mean, or should mean to the future of law enforcement. On top of the expert prediction that fMRI-based lie detection will be good enough to be presented in a Supreme Court hearing within five years, another expert says that in ten years brain tests will also be able to sort out the person who actually did a deed from anyone who was merely a witness. So we will need good answers, very soon, for an enormous package of questions.

In June 2008, a court in India found a twenty-four-year-old woman guilty of murder, based on a brain scan that the judge accepted as proof that she knew damning details of the crime. The system they used, and have also used in about seventy-five other cases to date, is based on reading brain waves with electrodes, looking for activity in memory-related areas. The judge wrote a lengthy post-trial opinion on his belief in the test's accuracy, though it has not yet been validated by independent study, or reported in a top scientific journal. American neuroscientists overwhelmingly see this development as premature and alarming, even incredible. But they also know how much pressure exists to latch on to a neuroscientific method of lie detection.

The many people who wish that the technology were bulletproof include judges, lawyers, police, homeland security personnel, parents, teachers, and all those people caught up in the justice system. According to the Justice Department, America's prison population rocketed from 1.1 million in 1990 to 2.1 million in 2006. By early 2008, nearly one American out of every one hundred was in prison.[1] This incredibly high rate of incarceration shows how tough the problem of crime is in the United States, and how far we're willing to go in the hopes of bringing it under control. Society's emphasis has swung heavily in recent years toward retribution rather than rehabilitation. California's "three strikes" legislation, which was approved in a 1994 ballot initiative, was an important moment in revealing that trend. Then came our post-9/11 fervor to identify spies, traitors, and terrorists. It is easy to see why the CIA and other intelligence agencies underwrite so much of the research in neuroscientific truth detection.

Many researchers hope that as neurolaw emerges, it will shift emphasis away from punishment. They believe we'll evolve toward "therapeutic justice" as we understand more about how brain disorders might cause criminal behavior, and about better ways to predict and prevent criminality.

Advances in neuroscience will have to fit in with three famous legal precedents. One, called the *Daubert* criteria, defines the requirements many U.S. courts use to determine whether scientific evidence can be admitted in a case. The Supreme Court made the *Daubert* criteria a precedent in a 1993 judgment that involved some truly wrenching emotions and heartbreaking circumstances but pivoted on a technical issue.

The essence of the *Daubert* criteria is that expert scientific opinion must be based on findings that have been published in scientific journals and have been subject to review by other scientists.

When Jason Daubert was born in 1974, he was missing one of the lower bones in his right arm. There were only two fingers on his right hand. His mother had taken Bendectin, an antinausea medication intended for pregnant women, while she carried him. Bendectin was introduced in 1957 by Merrell Dow Pharmaceuticals, a subsidiary of Dow Chemical, and was used by some 33 million women over the next twenty-five years. Its usage continued despite the fact that Merrell Dow was hit by hundreds of birth-defect lawsuits in the same year the drug was released.

Jason Daubert may well have been a victim of Bendectin. That was the opinion held by some highly respected scientists who gave testimony on his behalf. But, with irony that's painful to contemplate, Daubert lost the case at the same time his name was given to a legal standard.

What went wrong was a technicality. No one had yet published a study fully documenting the problems associated with Bendectin. The experts testifying on Daubert's behalf had to review data from several studies, each of which had been published in established journals and subject to peer review. They had to then compile the findings of these various studies and recalculate their data in order to make a presentation. Even though all the data was gathered from sources that met the criteria that from then on would hold Daubert's name, it was gathered into a brand-new document. This document, then, had not been published in a scientific journal. It was scientifically valid, but it didn't count.

The precedent set up by this case means neuroscientific advances won't be acceptable in the highest courts until they've been validated by the conventional process of scientific publication and review. That part of the Daubert story, at least, is reassuring.

The second legal precedent neuroscience will have to negotiate with was set in *Frye v. United States* in 1923, and clarified further in 1975 by Rule 702.

James Alfonso Frye was on trial for murder. The prosecution wanted to bring in evidence based on the work of a Harvard-trained psychologist named William Moulton Marston. A paper Marston had published six years earlier held that lies could be detected by tracking fluctuations in systolic blood pressure. In a 1911 report on Marston's work, the *New York Times* showed the same kind of heightened expectation that neuroscientific lie detection has put into today's legal atmosphere. That story predicted in the future "there will be no jury, no horde of detectives and witnesses, no charges and countercharges. . . . The

State will merely submit all suspects in the case to the tests of scientific instruments."

What Marston had invented was an early form of the lie-detection technology that still stands as the gold standard today, even though it has been proven so many times to be fool's gold.

The Frye test simply says that scientific evidence has to be based on a theory that the scientific community generally accepts. Three and a half decades before Rod Serling made *The Twilight Zone* the title of his famous TV series, judges in the case wrote this: "Just when a scientific principle or discovery crosses the line between experimental and demonstrable stages is difficult to define. Somewhere in this twilight zone . . . the thing from which the deduction is made must be sufficiently established to have gained general acceptance." In their opinion, the lie detector wasn't yet in the general-acceptance category.

Rule 702 was adopted to let in scientific testimony—even when it is not based on generally accepted theories—in cases where it will help judges and juries to better understand the evidence or reach conclusions. Under Rule 702, the testimony simply has to come from a qualified expert witness and be deemed relevant.

The third legalistic touchstone, and by far the oldest, dates to an English court's ruling in 1843. Daniel M'Naughten (sometimes spelled "McNaughton") had tried in 1812 to assassinate Robert Peel, then the British prime minister. He mistakenly killed another man.

The essence of the M'Naughten verdict was that the murderer was insane. Later, though, evidence turned up that M'Naughten might have been just a pretty good actor. What makes this a classic piece in legal scholarship is a question that still haunts judges and juries: When can we rightly say that a person is not responsible for his or her actions?

M'Naughten reportedly held a number of paranoid delusions, including a belief that both the pope and the British government were conspiring against him. He was a woodworker by trade, but he had previously been both an actor and a medical student. Glaswegian library records showed he had been reading about insanity just prior to the assassination attempt. It seems possible that he may have been faking to avoid a death sentence, but M'Naughten spent the rest of his life incarcerated in a mental hospital. His name ever since the 1843 ruling has been attached to the concept of insanity as a defense.

A panel from the House of Lords later reviewed his case at the request of

Queen Victoria. Twelve judges from common-law courts testified, and they helped the House of Lords refine the M'Naughten concept, establishing that insanity can be used for a defense only if the accused person is so mentally incompetent as to either not realize what he or she was doing, or not understand that it was wrong.

This was the first legal codification of an insanity defense. The idea itself has long roots, though, winding back to the Roman Empire and into classic Greek mythology. Hercules got off the hook for killing his own family, as well as the population of a whole village, because Hera, queen of the gods, had placed him under a spell. There was a catch to this insanity defense, though. Hercules had to cleanse himself by performing twelve great labors. That is why people still commonly say that a job that's going to be exceptionally tough will require "a Herculean effort." What we now call mythology was then a belief system. Both Greek and Roman law of ancient times actually did take mental disability into account in criminal procedures.

While we wait on neuroscience to give us something head and shoulders better, use of the polygraph lie detector remains widespread. Since Marston's time it has scientifically measured one thing: how nervous the test taker feels.

If it bothers you to lie, goes the theory behind the polygraph, pen-tipped needles scritching over a moving sheet of graph paper will register increases in various physical phenomena that the machine can measure on your skin, such as blood pressure and electrical conductance. It's possible that anyone who has had negative or intimidating experiences with authority figures is a strong candidate for a false positive, no matter how fine the content of her character might be. Anyone who isn't troubled much by telling a lie, or anyone who simply believes he is really telling the truth even when he is not, will probably pass. And if you are scheduled for a polygraph test but you're not yet a polished liar, just search the Internet. More than seventeen thousand sources of information are ready to coach you on how to beat the lie detector.

According to a 2003 report from the National Academy of Sciences, close to half of the research that's been presented to back the use of polygraphs is of low scientific quality, and the most that should be expected of these widely used "lie detectors" is to be right slightly more than half of the time.[2] What the polygraph really reveals best is a pair of truths we should never stop believing. One is our desire for a reliable way to know whether or not we're being fooled is so strong, it can lead us to questionable decisions. The other is how dangerous it is to fool ourselves by placing trust in a technology that doesn't really deliver.

In 1988, Congress passed the Employee Polygraph Protection Act (EPPA). It states that employers can't use lie detector tests, either before or after hiring someone, except in certain specific instances. Neither can they "discharge, discipline, or discriminate" against anyone who refuses to take a test. Further, they must keep an EPPA poster conspicuous in the workplace. A spur to the bill's passage was a 1987 lawsuit that the American Civil Liberties Union filed on behalf of state employees in North Carolina. Until EPPA, they found themselves routinely tested with such questions as "Who was the last child that got you sexy?" and "When was the last time you unintentionally exposed yourself after drinking?"

Loaded questions like those might incriminate anyone who possesses a normal cringe factor, triggered by an innate sense of decency.

At the opposite extreme, sociopathic personalities breeze through the stress of telling lies. They may even feel astoundingly positive about what they've accomplished. A man who created murderous havoc for several years in the state of Washington, for example, went door to door for his Pentecostal church yet also pursued prostitutes, many of whom he strangled.

Gary Ridgway compared his relationship to hookers to that of a junkie and drugs. He was long suspected of being the Green River Killer, a serial murderer with a career nearly two decades long, who terrorized an area around Seattle and Tacoma. Most of his victims were either prostitutes or runaway teenage girls whom he picked up when they were hitchhiking. Ridgway was brought in as a suspect in 1984 and given a polygraph test. He passed it and was released.

Eventually, DNA evidence led to his rearrest in late November 2001, seventeen years after he had beaten the lie detector. He was convicted of the murders of seven women, then confessed to another forty-one killings. Many believe the true count was higher.

Recent history is littered with equally horrific polygraph failures.

In 1995, when he was director of the CIA, William Casey released a statement that said, "I regret that I cannot discuss in public more detail about the actual damage done by Aldrich Ames. To do so would compound that damage by confirming to the Russians the extent of the damage and permit them to evaluate the success and failures of their activities. That I cannot do."[3]

Ames had been head of the CIA's counterespionage service. From that post he served Russia effectively enough to carve a niche as one of the worst spies in United States history. In 1985, for example, he disclosed the names of several American agents. At least nine were executed soon after. For just one of his many KGB assignments, he handed over enough documents to create a twenty-foot stack. Being so high in the chain of command, he had access in all directions. Ames was suspected for a while of being a mole, but he passed a polygraph test and continued his lethal undermining work for many years more.

Two different approaches to brain-based lie detection are the main competitors now for research funding and practical application in courts and beyond. Both rely on finding activity in regions of your brain where memories go to stay, and both work essentially like this: Suppose you kill somebody, with no witnesses to the crime besides yourself. A photo or verbal description of something only the guilty party would recognize, say a picture of the victim after the crime, would almost certainly make memory-encoding regions in your brain light up like neon at a casino.

One of these methods for tracing guilty responses is called "brain fingerprinting." The name was devised by the inventor of the technique, Dr. Lawrence Farwell, who has gone on to devote tremendous energy to finding practical uses, through a Seattle company that he set up, known as Brain Fingerprinting Laboratories. He's not the only one who believes the potentials are great. His research work has been extensively funded by the CIA and tested by the FBI and U.S. Navy.

His method uses electrodes held in place by a headband, waiting on a subject's scalp to detect a certain spike in electroencephalographic signals from the brain. This particular upshot of electrical activity is known as the P300 MERMER, short for Memory and Encoding Related Multifaceted Electroencephalographic Response. The stimulus could be something innocent, like a familiar face, or something sinister, like a murder scene.

To date, no one has been shown to have conscious control over his or her P300 response. However, that doesn't make brain fingerprinting infallible despite the recent adoption in India's justice system of a method that's heavily based on Farwell's work.

J. Peter Rosenfeld, clinical psychologist and neurobiology professor at Northwestern University, published a 2004 paper in which he detailed how a test subject could use easily learned mental tricks to derail the accuracy of Farwell's method for brain fingerprinting. Rosenfeld is working on a method that combines various indicators, including pupil size, pulse rate, and body temperature changes, in addition to the EEG data that Farwell's method tracks. He has tested only with volunteer subjects, not yet with suspected felons or spies. Rosenfeld claims 90 to 100 percent accuracy.

The other approach, which uses fMRI, is based on something called the Guilty Knowledge Test. Daniel Langleben of the University of Pennsylvania published his early work with this test in 2001. Years of studying the mental processes of heroin addicts supplied Langleben with a huge number of intriguing questions about impulse control and truth telling. Drug addicts are often superbly skillful and well-practiced liars.

In simplest terms, the Guilty Knowledge Test requires asking a subject several questions. Some of them are random and innocuous. Some of them can be answered only by a person who has highly specific knowledge. No matter what the subject says when answering the question, the fact that it makes him or her confront "guilty knowledge" is supposed to cause vivid brain activations.

There are some good reasons to believe that Langleben is working on solid scientific ground, even if we'll have to wait to be certain about his theory. Telling lies is harder work than telling the truth. More brain areas have to kick in. According to one recent study, seven parts of our brain pool their efforts when we tell the truth. When we're faking, fourteen areas have to come online. The frontal lobe carries the greatest strain, holding the truth in check while also framing up the "reality" that will take its place.[4]

Langleben's 2001 report stirred tremendous interest by revealing which areas of the brain go to work, as shown by fMRI imaging, whenever we take on the job of suppressing the truth and making up lies. Unlike Farwell, Langleben has stayed focused on new research, publishing a number of significant papers in recent years. They include "True Lies: Delusions and Lie-Detection Technology," a 2006 article in the *Journal of Psychiatry and Law*, in which he reviewed advances in lie-detection technology and suggested what additional things scientists must learn in order to make it reliable.

But even though Langleben isn't focused on the commercial possibilities of his findings, many entrepreneurs are, and with high hopes. Some are already selling expensive services; others are waiting until they can reliably generate accuracy scores of 95 percent or better.

Farwell promotes the results of his lab with zeal. He has sometimes cast his input in legal proceedings as pivotal when it was really a minor element, even though a radical one.

This doesn't impugn Farwell's scientific judgment, but it does cast a shadow. Of course, every business has the right and even the outright need to promote itself. But when that business is about decreeing innocence or guilt, whether it involves a single person's fate or that of hundreds of people, standards have to be the highest humanly possible. Billions of dollars are waiting to be made by credible neuroscientific lie detection. Just as in the story about Abraham Lincoln having a guard take his would-be corrupters out of the White House, the tension within neuroscience circles going after truth-detection methods, the push and pull between scruples and marketplace potential, may well be ratcheted up higher than most people can bear.

EEG is used in some very serious studies, but many researchers prefer the newer technology of fMRI. Stanford's Brian Knutson, the founder of a field of study called neuroeconomics, whom we'll meet at length later, says that attempting to judge what someone is thinking by reading an EEG can be compared to standing outside a baseball stadium during a game, trying to picture what's happening with the players by listening to the rise and fall of the crowd's roars.

In early brain fingerprint tests, Farwell showed subjects a series of seven-digit numbers. They eventually saw, randomly placed in the series, their own telephone numbers. Only the sight of their own number produced a spike in their P300. "The accuracy rate so far has been 100 percent," Farwell told a reporter for his hometown paper, the *Fairfield Ledger* of Fairfield, Iowa, adding, "All scientists know nothing is ever 100 percent, so I don't tout it as 100 percent accurate technology, but I do have high statistical confidence in it."

Evidence from brain fingerprinting isn't considered conclusive enough, at present, to convict someone of a crime in the United States. But it can be introduced as evidence in court, and has already figured in some life-and-death courtroom dramas.

One of these involved a murder suspect from Guthrie, Oklahoma, with the star-crossed surname of Slaughter. The crime, committed in 1991, was exceptionally grisly. According to prosecutors, Jimmy Ray Slaughter's girlfriend, Melodie Wuertz, twenty-nine, was stabbed in the chest, then shot, leaving her alive but paralyzed as Jessica Rae Wuertz, the eleven-month-old girl to whom Slaughter was father, was murdered by being shot in the head. The killer then finished off Melodie Wuertz, mutilating her with a knife, possibly attempting to carve the letter R into her flesh.

Farwell gave Slaughter a brain fingerprinting test in 1994. He said his test reliably indicated that Jimmy Ray Slaughter had no knowledge of the crime scene. Assistant Attorney General Seth Branham called the technology "junk science," but Slaughter won a stay of execution so an appeal could be considered. It was only a delay. What happened while it elapsed makes one realize how much clearer justice could be—and perhaps will be—if neuroscience can deliver reliable truth detection.

Ultimately, a Court of Criminal Appeals judge ruled that Slaughter "was not entitled to successive post-conviction review of claim that new evidence of brain fingerprinting would demonstrate actual innocence." So the brain fingerprinting test was not overturned, but simply ignored. Procedural standards made it a victim of bad timing. At the same time, the defense was blocked from introducing new DNA evidence that may have supported Slaughter's claim of innocence.

Meanwhile, according to a news report, the man who had initially been lead investigator on the case claimed he was removed from the case because he believed the investigation wasn't being properly handled. It turned out that there was another man, someone who also had a sexual relationship with Wuertz, who had disappeared a few days after the crime. His alibi was proven to be false, yet he was excluded from consideration as a suspect.

On March 15, 2005, Slaughter had a last meal of fried chicken and mashed potatoes, followed by an apple pie and a pint of cherry ice cream. Then he was strapped to a gurney in the death chamber. He told all those present, "I've been accused of murder and it's not true. It was a lie from the beginning. May God have mercy on your souls." Then he was given a lethal injection.

In 1999 Farwell was called in on another thorny case. He tested James B. Grinder, a suspect who had previously confessed to the 1984 rape and murder of Julie Helton in Macon, Missouri. Grinder was already in prison for another crime, and due to be tried in three more murder cases. He had confessed, but had also clouded his confession by telling contradictory stories at various times.

Farwell's test results verified that Grinder was familiar with the crime scene. When Grinder was told what his EEG confirmed, he struck a deal with prosecutors: life in prison with no possibility of parole.

According to Brain Fingerprinting Laboratories, "Brain Fingerprinting testing does not measure guilt or innocence, and nor does it measure participation or non-participation in a crime. It simply detects the presence or absence of information stored in the brain."

Farwell has been on *60 Minutes*, Fox News, *48 Hours*, ABC's *World News*, the *CBS Evening News*, CNN Headline News, and the Discovery Channel. The *New York Times* and *U.S. News and World Report* have run brain fingerprinting stories. The *60 Minutes* segment ran in December 2001. In it, Mike Wallace discussed the unsettling case of Terry Harrington, an Iowa teenager found guilty in 1978 of the murder of a retired police officer.

Farwell gave Harrington a brain fingerprinting test in April 2000. He testified that Harrington's brain did not contain memories of the crime scene but did contain memories of the events Harrington had described in his alibi.

The *Daubert* criteria are not required in Iowa district courts, but a district court judge named Tim O'Grady, in a hearing to decide whether the Terry Harrington case ought to be retried, ruled that brain fingerprinting met the *Daubert* criteria and could be admitted as evidence. Later, another Iowa district court ruling denied Harrington's petition for a new trial. The Iowa Supreme Court then reversed that ruling on constitutional grounds. When prosecutors lost that procedural round, they gave up and set Harrington free.

Farwell's potential testimony wasn't the only thing that convinced the prosecutors to concede. More pieces of exculpatory evidence emerged at about the same time. It turned out that there were eight different police reports that could have helped Harrington's case, but they had been withheld from his defense attorneys. Further, a witness against Harrington withdrew his earlier testimony, saying that he had been intimidated into giving it.

Although there were several factors that pushed the prosecutors to concede, the Harrington case created enormous interest in Farwell's work.

Langleben has also been in the popular media. In a study published in November 2001 and reported on in the *New York Times* one month later, Langleben and his team gave each of eighteen test subjects a randomly chosen playing card. The subjects were told to lie about the card they were holding, then were placed in a scanner and shown a series of cards. Each time a card was shown, a subject was asked if it matched the one he had been given. Therefore, he had to tell a lie at some point. When he did, a particularly important brain region (called the anterior cingulate gyrus) became more active in the fMRI images. This region is known to play a role in what are called the brain's "executive functions," including blocking one possible response in favor of another.

These early tests distinguished lies from truth with an accuracy rate of 77 percent. That wouldn't win a nod from the chief justice, but it was enough to inspire some serious commercial thinking.

The entrepreneur Joel Huizenga estimates the market for accurate lie detection might be $36 billion annually. He launched the jauntily titled business No Lie MRI to fill that need and, hopefully, cash those checks through commercial application of the Guilty Knowledge Test. One of Huizenga's biggest investors is Alex Hart, former CEO of MasterCard International, now a management consultant to No Lie MRI. Terrence Sejnowski, director of the Cricks-Jacobs Center for theoretical and computational biology at the Salk Institute in San Diego, is one of Huizenga's four scientific advisers.

Sejnowski is best known for work delving into the electrical and chemical changes that occur along with learning. Supercomputers process the data from his studies, attempting to prove how nerve cells in the brain do their jobs. Sejnowski is more interested in applications like the treatment of Alzheimer's than he is in lie detection, but he asserts—and most experts would agree—that methods used by No Lie MRI work better than the standard polygraph.

All of which explains why Huizenga is bullish on the future of No Lie MRI. "We're at the beginning of a technology that's going to become more and more mature," Huizenga told a reporter for *USA Today*. "But right now, we can offer [customers] a chance to show they are telling the truth with a scientific basis and a high degree of accuracy. That's something they haven't been able to get before."[5]

I met Huizenga at a 2007 investor conference in San Francisco. He is scientifically trained, with bachelor's and master's degrees in molecular biology, as well as an MBA. Most of his career has been a search for the entrepreneurial opportunities that could be discovered in the aftermath of scientific advances. He is still president and CEO of ISCHEM, his first start-up, which uses MRI to search circulatory systems for plaque that could result in heart disease.

Huizenga told me there are two pillars supporting the No Lie MRI business plan. First, a number of hospitals, universities, and private labs have expensive MRI scanners. When he tells them that renting their equipment's downtime could be a great source of income, smiles emerge. Second, the $10,000-per-session pricing of No Lie MRI testing just might look like a huge bargain to someone who wants to sidestep the financial steamrollering of a costly legal battle.

No Lie MRI has a competitor in Cephos, a Massachusetts deception-detection firm headed by Steven Laken. The Cephos Web site says the company "uses the latest advances in medical imaging to peer inside the inner workings of the brain during deception." Laken says that the company is still ironing out kinks and hasn't yet set prices. Its current policy is that the tools are still in development, but Laken believes that refinements they're about to test will soon boost their accuracy rate, now a reported 90 percent, up to 95. Laken expects to be offering tests very soon but cautions would-be customers: More testing must be completed before any commercial firm can claim that it's offering a valid truth-detection test.

His business plan has its foundation built on patents that were applied for by the Medical University of South Carolina in 2002 and later licensed to Cephos. His target market is the huge number of cases annually that hinge on a "he said–she said" situation, where proving someone's veracity might spell victory.

Laken sees lawyers and their clients as being the first buyers Cephos will attract, but he also looks beyond to demand from the defense and homeland security sectors. Certainly, fMRI-based lie detection, which doesn't involve waterboarding, electroshock, sleep deprivation, or physical threats, would improve the public's perception of a government that has recently bent the Geneva Conventions into a pretzel.

Laken is a scientist as well as an entrepreneur. His credentials include a doctorate from Johns Hopkins in cellular and molecular medicine. But even as fine an academic pedigree as that doesn't calm the concerns of skeptics like Hank Greely, a leading authority on neuroethics and a professor of law at Stanford. "I want proof before this gets used," Greely said. "And proof is not three studies of forty college students telling lies about whether or not they are holding the three of spades."[6]

Greely has an important point. For example, we need to know if different brain mechanisms may be at work when the stakes of the game are changed.

Maybe the potential consequences of the lie make a difference. John Gabrieli of MIT shared that concern with me one afternoon when I visited his campus. "The biggest problem is relating an experiment to real-life situations in which lying really matters. Still," he allowed, "greater precision may turn out to be useful."

Until their accuracy rates get nearer to 100 percent, neither Cephos nor No Lie MRI will be embraced by a majority of neuroscientists.

However, despite wariness and outright resistance, it seems likely that neuroscience will help us update the M'Naughten precedent of 1843.

Consider the fact that a century or so ago there were essentially just two different diagnoses for mental illness: feeblemindedness or lunacy. As therapies evolved, so did a good reason for delineating mental illnesses more finely. Similarly, we now have tools for looking inside a living, functioning brain, and they are beginning to help us develop more accurate ideas about when people can or cannot control their own behavior. The same dynamic that has given us so many different labels for the varieties of mental illness will probably drive the evolution of better, more finely tuned legal distinctions about appropriate degrees of cause and blame for antisocial acts.

If the M'Naughten precedent does manage to evolve, much credit—or blame—may be ascribed to Adrian Raine, an Englishman who became a Californian.

After he earned his doctorate in psychology from York University, Raine went to work as prison psychologist for two top-security institutions while simultaneously lecturing in behavioral sciences at Nottingham University. In 1986, he became director and principal investigator for the Mauritius Child Health Project.

Typically, antisocial behavior passes through families. The Mauritius project is an enormously ambitious cross-generational study aiming to understand how it gets handed down. The project was begun by testing 1,795 three-year-old girls and boys on the tropical island nation of Mauritius, in the southwest of the Indian Ocean, off the coast of Africa and about 560 miles east of Madagascar. Back when Mark Twain visited Mauritius he decided it was the first draft for heaven, the original from which paradise was copied.

After extensive testing of both the children and their parents, one hundred children were chosen for a two-year program of enhanced nutrition, exercise, and education-enrichment intervention, which they would receive between the ages of three and five. Those children are now in their thirties. At age eleven they showed increased physiological arousal and attention, relative to children who received no intervention. At age seventeen they showed less misconduct than their peers. The study is continuing to test the now-grown children, as well as their spouses and their own children. Researchers hope to find that richer socialization can help break antisocial behaviors within a generation.

The concerns around which he launched the Mauritius project still drive Raine's work today, but now he uses neuroscience to search for answers. Raine came to the University of Southern California in 1987 and since then has won several awards from the National Institute of Mental Health as well as one from USC in 2003 for creativity in research.

The neurobiology of violence is a central focus of his research, along with antisocial behavior, alcoholism, and schizophrenia, from both the genetic and behavioral perspectives. He studies pretty much anything that could come under the heading of "the criminal brain" and tries to understand how various kinds of antisocial thoughts and actions are driven. As with the Mauritius project, there's an ambition to establish some fundamental truths that can be used to help straighten out social problems and the people who cause them. The quieter corners of neuroscience are just as important to watch as the stories like work on truth detection, which feed into the popular media more easily. The quieter work of Raine and many other neuroscientists may cause the most dramatic developments of neurolaw to take place outside of courts and interrogation rooms.

For example, Raine used neuroimaging to trace the mental activation patterns shown by people who have been guilty of battering a spouse. What he found proved that the batterers were impelled by a very specific kind of emotional pain, the fear of abandonment.

When people who batter their spouses are shown enactments of abandonment, in which a battered spouse announces his or her independence and leaves, the areas of their brains associated with anxiety and anger fire up heavily. This has huge implications for treatment, and for lifelong prevention through richer socialization.

This research out of USC is just one part of a growing body of evidence that is going to force us to rethink our ideas about crime and punishment. From one point of view, it can be seen as scientific proof that we need a more compassionate and nurturing society, and that investments in better socialization will pay off in a big way, breaking the chain of cause and effect, making us all safer. From another perspective, it gives criminals more opportunities for playing the angles.

Looking forward, we can expect at least two other key things to happen. Scientific proof of the fact that criminal behavior is very often tied to brain abnormalities and diminishments will inspire more defendants to try scientifically based variations on a defensive stance of "the devil made me do it." This, in turn, will pressure the justice system to pin down a reasonably definite and workable standard for deciding when individuals are, or are not, truly deserving of punishment.

The brain is the control room for decision making. Now that we're learning more about the conditions that impair a person's decision-making ability, and we can actually see in an fMRI scanner whether a person is, in effect, being pushed in wrong directions by faulty wiring, we may feel more like helping the afflicted people to get better. For example, between 55 and 90 percent of people arrested for felonies in our largest cities test positive for controlled substances or alcohol. Thirty percent of state prisoners and 40 percent of federal prisoners are locked up on drug-related charges. Think of drug taking as a self-administered, and poorly chosen, regimen to fight the inner turmoil of mental illness. This assumes, of course, that the addicts don't necessarily recognize that's what they're doing. More likely, they just know that they don't feel well and seek to get relief from their painful emotions.

Meanwhile, experts estimate that about one out of every four violent criminals in prison has diminished brain function of some kind or other, whether retardation or physical damage. It is very common for an autopsy of a convicted killer to show extensive frontal lobe damage from concussions. There have been many tragic stories over the past few years in newspaper sports sections about ex–NFL players who have died violently or have been incapacitated by lingering brain damage from years of slamming their massive bodies into equally massive opponents with all the force and focus they can generate.

Imagine what could happen if we figured out how to update the M'Naughten precedent appropriately. Combine that thought with another one. What if neuroscientists in the field of medicine, for example, also invented better therapies for people with violent or antisocial tendencies? Blending neuroscience knowledge into our legal system could inspire a fact-based belief in forgiveness and compassion, along with more interest in rehabilitation and less in severe punishment. This may sound like a concept pulled out of the clouds, but right now we're warehousing an ever-growing number of people in a prison system that keeps on bursting at the seams no matter how many new facilities we build. Instead of funding a system that serves as postgrad training for career criminals, we may find ways for brain science to help us evolve more effective solutions to crime and rehabilitation.

Of course, no one can really predict the future, and we may be surprised by the actual developments that eventually result from neurolaw. However, right now, here is where we should be looking. Both neuroscience and psychology have made enormous progress in recent years. But there isn't enough interaction yet between the two disciplines. A massive gap remains to be filled. We still have no well-substantiated technology that will allow us to read minds, detect deception, or determine guilt or innocence with brain images. The experimental literature consists of only sixteen studies, with results that remain inconsistent and unreliable. Commercial enterprises offering brain imaging as a lie-detection process are at best premature, and are likely to fail for purely scientific reasons. Until the technology is really solid, these enterprises are going to be promoting the same kind of pseudoscience that has let the polygraph hang around on the fringes of our judicial system for the past century.

Stamping out the attractive and popular idea that we can read minds with brain imaging is as difficult as killing vampires, but we need to drive a few wooden stakes in by insisting on the best possible scientific scrutiny. We are going to live for a while in a period of uncertainty and confusion about neurolaw, at least until accurate, validated, peer-reviewed techniques arise.

But that still leaves us plenty to hope for.

For one thing, technologies can grow in unpredictable ways, and barriers sometimes fall down on their own schedule, not the one experts saw coming. Nicholas Negroponte's excellent 1995 book, *Being Digital*, points out that after our ability to compress and decompress computer data grew a lot faster than expected, the digital era began rolling out centuries sooner than some experts had believed it would. Little advances in one area can have multiplying effects in several others. There are so many exceptional minds working in neuroscience that it's foolish to rule out the possibility of lightning striking.

For another, there are also some incredible minds working right now on showing neuroscience and the wheels of the legal system how they might mesh to improve the generation and delivery of justice.

One of the most amazing women I've ever met, Margaret Gruter, is someone we have to thank for that. Dr. Gruter died in 2003 at the age of eighty-four. I had met her just a few months before that, at one of the most important conferences I've ever attended.

When Hitler was building his regime, Gruter was a young German woman whom the Nazi authorities classified as "unreliable." She saw large-scale sociopathic behavior firsthand, persecution and destruction of individual lives, sanctioned and carried out by those same authorities. Seeing how a society could misshape lives made her wonder if those gears could ever be reversed. That became the theme of her life's work. Gruter's training and personal experiences impelled her toward concepts of fairness, justice, and ethics, and to believe that science and the law should work together. After an extensive undergraduate program in the humanities, Gruter went to law school at the University of Heidelberg, where she took a doctor of jurisprudence degree.

She immigrated to the United States in 1951, first assisting her husband in his medical practice in rural Ohio, later becoming administrator of a medical facility for the developmentally disabled. Seeing how those patients struggled against the limitations decreed by their neurobiology, even in the most nurturing environment that could be provided, she wanted to see the best of available science used in understanding their conditions.

She and her husband moved to California in 1969, and she enrolled at Stanford's law school. Her interest in how social values get formed led to several deep conversations with Dr. Jane Goodall, and in 1972 to meeting with Dr. Konrad Lorenz, the Nobel laureate famous for his studies of animal societies. He encouraged her to keep looking for ways to connect law and science.

In 1981 she founded the Gruter Institute to further the concept that legal systems must work to help people change their lives, and that one way to make that happen was for people in the legal profession to understand scientifically how law influences behavior.

Annual conferences of the Gruter Institute have always been famous as amazing assemblages of intellectual power and humanistic values, where information flows between biologists, including neuroscientists, and legal-system professionals. I first attended in 2003.

I was in the process of writing this book and had been learning all I could from recently published scientific papers. But at the conference, held at PlumpJack, a Squaw Valley ski resort in California's Sierra Nevada mountains, I suddenly came face-to-face with many of the most brilliant researchers in their fields. I particularly remember an incredible dinner session, over delicious roast duck that I barely paid attention to, while the topic of free will surged back and forth across the table between Paul Zak, Paul Glimcher, Howard Fields, Oliver Goodenough, Margaret Gruter, Kevin McCabe, and Morris Hoffman. Many of those names will pop up in significant ways in various chapters ahead. It's hard to express fully how inspired I was that evening, submerged in their tremendously open conversations about all the ways in which brain research could totally reshape our future lives. I felt like a mostly self-taught musician who suddenly and unexpectedly got invited into a jam session with Alicia Keys, Eric Clapton, Bono, and Bruce Springsteen. We talked about randomness, arrows in time, coin flipping, converging utility maps, selective evolutionary tendencies, dopamine neurons, motivation, oxytocin, synesthesia, and whales. It was amazingly intense, yet collegial. No one acted like he or she had all the answers, even within their own fields, let alone the many other patches of study where neuroscience is expanding knowledge and horizons. But everyone dealt out the smartest, most provocative questions.

At PlumpJack I began to realize something that has shown up again in other conferences several times since. No matter how brilliant these people were, they were so committed to keeping up with their own area of study that they didn't have enough time to build out their awareness of what was happening in other disciplines. Meanwhile, I was taking in everything that I could, wondering how to synthesize all the various things I was learning so I might be able to push it far into the future, and see what it might look like through Time's Telescope.

In 2007, four years after Gruter's passing, the foundation bearing her name became the key player in education and outreach aspects of a $10 million grant given by the John D. and Catherine T. MacArthur Foundation, for the purpose of helping the U.S. legal system understand and integrate current developments in neuroscience. The Law and Neuroscience Project is the collective name of the entity, which will run for three years and—if MacArthur officials see promising results—probably carry on for several years more. Announcing the grant, Jonathon Fanton, who is president of the MacArthur Foundation, said, "Neuroscientists need to understand law, and lawyers need to understand neuroscience. It could have an impact on the legal system as dramatic as that of DNA testing."

I think that's an understatement.

The project is centered at the University of California, Santa Barbara (UCSB), and a UCSB psychology professor, Michael S. Gazzaniga, is its director and principal investigator. Former Supreme Court justice Sandra Day O'Connor is serving as honorary chair.

Gazzaniga is the author of a 2005 book, *The Ethical Brain: The Science of Our Moral Dilemmas*, and was a member of President Bush's Council on Bioethics. He now has scientists and legal scholars from more than two dozen universities nationwide working on different aspects of the project. They are divided into three main networks, each network investigating a single area.

Gazzaniga, together with Stanford's Hank Greely, will direct research involving diminished brains, that is, brains which by accident of birth or injury have significantly less capacity to know right from wrong or to control their owners' behavioral impulses.

The second network will research addiction and antisocial behavior. The third will focus on the neurobiology of decision making. Each network will meet at least three times per year, and all three networks will join together for at least one annual meeting, together with a conference for presenting research.

The most important first step in the project is to decide where the gaps are in our current scientific knowledge in order to quicken momentum around pressing neurolegal questions. The various answers that the networks shape around that question will set specific courses of action for the project, which will tap into the best scientific minds available whose research might fill those knowledge gaps.

Gazzaniga and his collaborators in the project started with a long list of vital questions competing for their focus. How does someone turn out to be a criminal? Is there a biological basis? What is different in the brains of people with self-destructive tendencies, like drug abusers and compulsive gamblers, compared to those of people with wealth who risk prison in order to gain a relatively slight bit more money? What are the neurological effects of being raised in poverty or in crime-stricken neighborhoods? Why does one person react to a provocative situation with reason and constructiveness, and another person in exactly the same situation turn violent? What goes on in our brains when we're trying to make crucial decisions within a group, such as a jury? And, the biggest question, can neuroscience eventually give us highly reliable detection of truth?

The project's leaders decided that out of all possible areas, their first major focus should be the theme of criminal responsibility, along with a closely connected theme, the responsibility of society toward criminals.

Criminal responsibility is a solid place to start. It is an area of law that is fairly well defined, so there's some structure in place to work with, and some important lines of debate already exist. For one example, where exactly should society draw lines saying what constitutes a condition that's beyond individual control? These decisions have to be exceptionally well made, so the legal system needs the best possible input.

A little over forty years ago, a tragedy happened on the campus of the University of Texas in Austin. It riveted the whole nation's attention for a long time. Unfortunately, it's an event that has been echoed frequently in the years since, and it illustrates how deep the complexities are within the subject of free will and individual choice.

On August 2, 1966, an ex-marine and ex-student named Charles Whitman carried a passel of weaponry up to an observation deck on the university's thirty-two-story administration building. He had already murdered his wife and his mother. His arsenal was a sawed-off shotgun, a hunting rifle with a scope, another rifle, an M1 carbine, and three handguns. The deck was a sniper's paradise. It took hours before police were able to get a clear shot at Whitman. By that time he had killed fourteen people and wounded thirty-one others.

Whitman had been honorably discharged from the Marine Corps after misconduct. He had been prescribed Valium by a campus doctor. That doctor had also referred Whitman to a psychiatrist, who found him to be "seething with rage." Not a surprising conclusion, since Whitman had admitted that he had felt an urge to go into the university's tower and "start shooting people with a deer rifle."

The night before he climbed into the tower and began killing at random, he began writing a suicide note: "I do not quite understand what compels me to type this letter. Perhaps it is to leave some vague reason for the actions I have recently performed. I do not understand myself these days. . . . However, lately (I cannot recall when it started) I have been a victim of many unusual and irrational thoughts."

After murdering his mother and his wife, Whitman added this: "I imagine it appears that I brutally killed both of my loved ones. I was only trying to do a quick and thorough job. . . . If my life insurance policy is valid please pay off my debts . . . donate the rest anonymously to a mental health foundation. Maybe research can prevent further tragedies of this type."

When the coroner did an autopsy, there was a cancerous tumor found in Whitman's brain, pressing against a region called the amygdala. That tiny piece of the brain, only about three-fourths of an inch in length, is a major center for processing emotions.

This was an incredible and sickening course of events. Because Whitman was able to stay holed up for such a long time, the sequence of deaths unfolded on national broadcasts. Each update pushed the question deeper into the public mind: How could anyone do this?

Today, if a psychiatrist were to interview a patient in a mental state anything like the one exhibited by Charles Whitman, that psychiatrist could order an immediate brain scan. MRI would reveal the impinging growth. If the tumor were operable, surgery would follow as instantly as possible.

In terms of the Law and Neuroscience Project's intended research, scientists have already found that a tumor of the same size and in the same general location could affect another person quite differently than Whitman's tumor affected him.

Whitman had a very high IQ, and he had been an Eagle Scout and a volunteer scoutmaster. But he had also physically abused his wife, just as his father had physically abused his mother. He had grown up with guns and hunting, and had extensive weapons training in the Marine Corps. Did his life experiences worsen the outcome of having a tumor compressing such a tremendously important part of his brain? What exactly happened to signals in his brain because of the impairment?

To distill some gigantic complexities into a single, simple question, when should the legal system accept "My brain made me do it" as a valid defense, and when should it not?

An argument that's often put forward is that guns don't kill people, people kill people. (The comedian Eddie Izzard says he agrees, but feels that the gun plays a very important role.) Therefore, we should set up different kinds and degrees of punishment depending on whether or not a person intended to commit a criminal act.

However, brain imaging studies indicate that our brains start producing their reactions before we are conscious of our intention. Free will has been debated in religion and philosophy for as long as those fields have existed, but psychology and neuroscience now indicate that it mostly exists in our imaginations. This is an unsettling piece of news to assimilate, but it may turn debate toward questions that can really be answered, thanks to brain scanning technology. Those answers could become important building blocks for legal policy, and help address other social issues.

Adding to the difficulties we face in figuring out intentionality, recent studies have shown that teenagers' brains are not fully developed in certain ways, particularly in the frontal lobes, which are crucial to reasoning skills, such as weighing the risk of actions and controlling impulses. This was pivotal to a Supreme Court challenge in 2005 in the case of *Roper v. Simmons.* Both the American Psychological Association and the American Medical Association filed briefs based on brain scan studies that showed adolescents have diminished brain capacity, compared to adults, and shouldn't have to face the death penalty.

However, before society can decide whether or not to weigh neural evidence, we all need full clarification of what neuroscience can do, and what it can't do, in pinpointing the causes of criminal behavior. So Gazzaniga and all his co-researchers have the lives of many in their hands.

"We must revise our notion of social responsibility," Gazzaniga told a Washington, D.C., audience at an October 2007 lecture entitled "Brains, Minds and Social Process" at the Carnegie Institution for Science.

Although the brain is what we use to make our decisions, Gazzaniga insisted in his talk, we don't actually control the decisions our brain makes, at least not to the extent that most people believe we do, and not to the degree that the law supposes we do. He ran down the long list of legal issues that neuroscience may help us resolve, like whether anyone can truly be not guilty by reason of insanity, how to accurately diagnose states of diminished consciousness, and the ways that trial evidence can be impacted by behavioral biases. These questions have a long history of controversy within our legal system, but he added that we are starting to find scientific answers. "These are all tough, tough questions," Gazzaniga said, building to his conclusion. "We're nowhere near being ready to answer them all."

Let's assume for a moment that neurotechnology advances at an accelerated rate over the next two decades, giving us extremely precise brain scanning technologies at low cost. When microchips made those same improvements—increasing power and plummeting cost—we got the necessary computing power to decipher DNA for proof of guilt or innocence. When we get a similar level of brain scanning power, and enough agreement about what the evidence found in brain scans means, neurotechnology will prove its ability to transform how legal systems operate.

We should expect the transformations to come in many different flavors, some of them bitter and some of them sweet. As the power to decode the brain accelerates, changes will sweep quickly across every society in a slightly different way. In more open and democratic societies, truth-detection systems will protect and free the innocent, a phenomenal improvement for those fortunate enough to prove their innocence with the new technological tools. But we can expect that closed and autocratic regimes will leverage the same technologies to silence dissent and enforce loyalty to the current leadership.

As we clarify our understanding of how brain diminishment—whether because of birth defect, injury, or disease—impacts individual behavior, we will be able to generate powerful new therapies. These interventions will not only transform how we treat mental illness; they will also reshape how justice is carried out. Instead of doing hard time behind bars, criminals of the future of justice may be sentenced to soft time under the influence of mind-altering drugs that annihilate addiction, restrict rage, and expand empathy. But this same knowledge of how to influence or control the human brain, if in the hands of unscrupulous people, will be used to eliminate memories, incite anger, and instigate violence.

When neurotechnology enables us to predict an inclination toward criminal behavior, some countries will scan their populations to head off danger before it begins. Those who are likely to get snared by addictions may eventually be vaccinated. Researchers are already developing vaccines for treatment of nicotine and cocaine addiction. They believe that these new medicines will prevent addictive substances from crossing from the bloodstream into the brain.

Brain privacy will become the civil rights issue of the twenty-first century. In many countries, legal and political frameworks will be developed to protect individuals against brain discrimination. In the United States, the recently enacted Genetic Information Nondiscrimination Act (GINA) took nearly a decade to pass, but it now bars employers and health insurers from discriminating against people on the basis of their genetic information. This law proves that transformative technologies—such as DNA testing—bring about sweeping legal changes. With GINA's passage, Americans can now take advantage of the tremendous potential of genetic research in promoting health without fearing that their own genetic information will be used against them. Following the same pattern, advances in neurotechnology will create a profound desire in society to protect our cognitive liberty as individuals. This could lead to a Brain Information Nondiscrimination Act whose protections, in turn, will further accelerate the use of neurotechnology for positive, helpful ends.

Neurotechnology will be extensively used in the courtroom, for many purposes. These will include determining bias, compelling the truth, and showing whether someone poses a risk of future criminality. We can expect truth detection to continue as a hot and leading issue, but we can be sure that neurolaw will roll out in several different directions in the not-so-distant future and inspire us to update not just the old M'Naughten precedent but literally thousands of the rules of law by which we live.

THREE

MARKETING TO THE MIND

We study theory in order to apply it, not for its own sake. —Ho Chi Minh

Much of a man's thinking is propaganda of his appetites. —Eric Hoffer

A quick note about your brain: It is generally organized into three parts. Your most newly evolved brain tissues are wrapped around the outside. Further beneath the surface are the two vintage structures, the mammalian brain and then the reptilian brain. These older parts are generally the sites for basic functions—breathing, circulating blood, digesting food, and so on. They also operate undetected by conscious thought. They are constantly making rapid choices for you based on immediate sense impressions, like the features and expressions on faces of people you see. The newer, close-to-surface brain parts draw from all of the various brain regions, weigh different inputs, and then act like a slowly deliberating committee. The older regions act like an impulsive dictator, firing commands, unimpeded by the need to seek balance.

Much of the time, your newer and more thoughtful brain parts don't realize they're being bossed by the older ones. Maybe the best way to connect with this fact is through a country tune I happened to hear one day. It was about giving in to temptation and getting into hot water, and the lyric was built around a catchphrase we all can relate to: "I know what I was feeling, but *what was I thinking?*"

That's how we humans are. We're frequently pretty mysterious creatures, even to ourselves. We experience subtle thoughts with powerful feelings, then powerful thoughts with subtle feelings, and we often don't know what to do about either of them. A lot of the time we don't even know what lit the sparks. Not only do we make choices without knowing what we were thinking, we reject options that clearly would have served us better.

If we puzzle ourselves, we throw the people trying to influence our choices of goods, services, or politicians into profound confusion. The American economy inspires $120 billion worth of advertising buys annually. More than $8 billion is spent per year for advertising market research in the United States, $1 billion of it for marketing focus groups.

The people behind those millions and billions are extremely hopeful that brain imaging can help them swing your choices in directions that they choose. Be assured that they're working on it. "Instead of hypotheses about what people think and feel," a Virgin Mobile USA marketing exec talking about brain imaging told the *New York Times* in March 2008, "you actually *see* what they think and feel."[1]

That annual billion spent on focus groups illustrates the core mystery that marketers want to penetrate. You can get people together and ask them to compare stuff, and you can tabulate which brands, or features, or flavors, they say they like the most. But most of the time people are just like that guy I heard singing on the radio: They don't really know their own minds. What people praise in the focus group, they may not be willing to trade precious dollars for later.

Brain imaging can show what is going on in our minds when we make choices even if we ourselves don't know. But the marketing industry isn't funding most of the broad, general studies, the kind that try to establish fundamental truths. Marketers are after specific applicability, secrets that can move their brands off the shelves before the competitors get wise.

It's a good thing to be wary of studies that come directly from marketers, or from anyone who has a direct economic stake in the findings. Several psychological studies have proven how easy it is for bias to sneak its way into our thoughts and distort the accuracy of the conclusions we draw. (For example, a scientist whose lab experiments are funded by Hilda's Tastee Soups Inc., even if he or she aims with heart and soul for unadulterated truth, may sooner or later produce studies that prove Hilda's Tastee Soups are unsurpassed.)

The fundamental and groundbreaking studies are generally paid for by government agencies, such as the National Institutes of Health (NIH), but money for cutting-edge science can be hard to line up. Neurotech research tends to be very cross-disciplinary, and traditional sources like the NIH have very specific categories and guidelines that were drafted for a different age. That means that researchers need to be at least as entrepreneurial about getting their funds as they are inventive in framing their scientific thoughts.

One of the pioneering researchers in brain imaging and financial decisions, Brian Knutson, whom you met briefly in the previous chapter, recently scored grants from two lesser-known sources: the National Association of Securities Dealers and the eBay Foundation.

Another top-gun researcher, Read Montague, whom you'll also meet a few pages ahead, does work that would fascinate military minds, but does not seek Defense Advanced Research Projects Agency (DARPA) or any other military funding. He's concerned that taking their money might lead to pressures that would distort his research. Nevertheless, he recently convinced his university to acquire three new MRI scanners, at about $3 million apiece, to combine with the two already on site. This unprecedented five-scanner cluster will make it possible to test how people's brains react in group situations, such as team building and consensus creation.

Money from government sources and from private charities is responsible for the foundational studies in neuromarketing. Individual companies are using neuroscience too, but for highly specific studies on certain fine points, like how to acoustically engineer a reassuringly solid *thunk* sound when a car door closes, an aural experience that may convince buyers that the whole car is very well made.

There already exists a host of emerging neuromarketing companies that offer corporate customers a chance to shine a neuroscientific light into the consumer's mind, in hopes of understanding how and why people choose specific products. They include San Francisco's EmSense, which measures brain activity "for a moment-by-moment analysis of how audiences respond to advertisements"; London's NeuroCo and NeuroSense; Berkeley's NeuroFocus, which is "applying the latest advances in neuroscience to the world of advertising and messaging"; Atlanta's BrightHouse; FKF Applied Research of Los Angeles, the self-proclaimed "Leader in NeuroMarketing"; and Boston-based Arnold Worldwide and Digitas.

Right now, most of what these firms do is testing ad concepts so companies can decide which versions of their pitch are most likely to win over the people they want to reach. Essentially, they're doing the kind of work usually done with focus groups, but in a more streamlined and possibly more cost-effective way. I'm betting that we can expect much more sophisticated uses of brain imaging to emerge soon. For example, directors of commercials in the near future will probably shape their final cut after consulting brain-imaging results.

Gerry Zaltman, an emeritus professor of business administration at the Harvard Business School, is considered the first researcher to use fMRI in marketing studies, circa 1999. Zaltman believes that fully 95 percent of our thinking occurs in the subconscious mind. Because of that, marketing succeeds best when it makes a psychological link with our deep brain areas. If our old brain parts decide a product will make us feel connected to a larger group or help us hook up with a desirable mate, we're going to want to buy it. So this kind of psychological link in many markets is far more important to a corporation's marketing strategy than actually making a superior product. In other words, being the most desired pays better than being the best.

Although Zaltman is credited with the first work in the field, the word "neuromarketing" was coined in 2002 by Ale Smidts, director of the Centre for Neuroeconomics at Erasmus University in Rotterdam in the Netherlands. Smidts started studying advertising because it's a rich source of material for understanding how people become persuaded. Studying the techniques and strategies of advertising, he believes, will improve our understanding of the human mind.

One of his first studies looked at endorsement, a long-established advertising gimmick of using an expert or a celebrity (or someone who is both) to convince people that the product being endorsed is what a cool person would want. In a 2006 presentation at the University of Michigan, Smidts shared imaging studies proving that even a *single* exposure to a combination of product and expert "leads to a long-lasting change in memory for an attitude towards the product."

Smidts hopes that people will understand that there's a difference between neuromarketing as a field of research, and neuromarketing as a tool kit for persuaders. For a whole lot of reasons, some of them totally valid, many people look at the advertising industry with a lot of nervous apprehension. A perfect example is in the 2005 movie *Thank You for Smoking*. Aaron Eckhart's character, Nick Naylor, is as charming as a golden retriever but as amoral as a hammerhead shark.

For at least some people who live by their skill in persuasion, deception and ruthlessness often are standard operating procedure. We fear their tactics because we realize that we're all vulnerable, even the brightest among us. The idea that these persuaders now have brain science on their side can intensify that fear.

Ralph Nader founded an organization called Commercial Alert in 1999 as a watchdog group to counteract commercialism. In 2003 Commercial Alert sent a letter to a federal agency, the Office for Human Research Protection (OHRP), saying that the use of medical technology such as brain imaging in marketing is wrong. The letter called for an investigation of Emory University, which was at the time allowing the Atlanta neuromarketing firm BrightHouse to use the university's brain imaging equipment for marketing research. Commercial Alert wanted to know whether federal guidelines on the use of human subjects in experiments were being violated. Its position is that enabling the growth of consumption would promote disease and suffering. Should the OHRP have decided to investigate, and subsequently have agreed with that stance, one of America's leading life science research universities would have lost federal research funds.

An outcome like that may seem far-fetched, but Nader's car safety book of 1965, *Unsafe at Any Speed*, had a huge effect on federal regulation of the auto industry, and his maverick presidential campaigns of 2000 and 2004 had the deeply ironic impact of putting antiregulation forces into power.

People in the persuasion business have been looking for help from science since long before Nader's time. In 1898, E. W. Scripture described a technique in his book *The New Psychology*. He called it "subliminal messaging." It consisted of flashing an image or a set of words so rapidly that the conscious mind couldn't register them, though the subconscious would.

In 1957 an adman named James Vicary claimed he'd tapped the power of subliminal messaging, and touched off a public controversy that kept running for years. Vicary said that during screenings of the movie *Picnic* he had sent rapid-flash messages at five-second intervals to the audience. They reportedly included "Drink Coca-Cola" and "Hungry? Eat popcorn," and they lasted just one–three hundredth of a second. Vicary said his technique made people head for the lobby and buy significantly more popcorn and other snacks.

Vicary eventually was revealed as being to marketing what the con man Frank Abingale—played by Leonardo DiCaprio in the 2002 movie *Catch Me If You Can*—was to the banks and airlines he bilked. Scientists working in laboratory settings tried to duplicate the experiments Vicary claimed he had done, but they never got comparable results. Eventually, Vicary admitted to falsifying his *Picnic* report and other purported studies. But for a long time, the technique Vicary called "subliminal advertising" took on a life of its own. It had the public stirred into outrage and the marketing community thrilled. Vicary enjoyed a run as its guru by founding the Subliminal Projection Company. Meanwhile, his bogus research fired up sales of a popular 1957 book called *The Hidden Persuaders*, by Vance Packard, a lengthy warning about how the public was being duped by advertising and marketing.

By 1958, subliminal advertising was banned in Australia and the United Kingdom, by television networks in the United States, and by the National Association of Broadcasters.

In late 1969, a researcher named Herbert Krugman began testing what happens within a person's mind when he or she watches television. His test involved attaching a single electrode to the back of each subject's head. Krugman concluded that TV viewing tended to shift people into a passive and receptive state, characterized by alpha waves emanating in the brain.

Krugman also pioneered the use of pupilometers, devices that measure changes in pupil size. When we believe something to be worth tuning into, our pupils automatically dilate. A similar technology, eye tracking, has also been used in hopes of creating more compelling ads. Eye tracking makes a record of the path a person's eyes blaze when assimilating a visual message. French scientists who began studying eye tracking in the 1890s determined that eyes typically bounce from one place to another, hopping and stopping, zigging and zagging, as if the brain wanted to find the quickest way to unlock the information. Research into eye tracking spread to Russia, where a psychologist named Alfred Yarbus studied the phenomenon and wrote *Eye Movements and Vision*, a 1965 book that was translated two years later and published in

America. Today, eye tracking has carved out a small but important niche in advertising design, helping researchers figure out where you focus, what attracts your attention first, and how you scan or read an advertisement, newspaper, or product package.

Galvanic skin response (GSR) is another scientific measurement that got advertisers' hopes up, particularly in the 1960s. It measures changes of electrical conductivity on skin surfaces, which are caused by emotional reactions. Researchers have used it to track emotional responses to brand names, background music, and advertisements. The same technology is behind the conventional lie detector, a machine that is none too trustworthy.

Another three-way collision of publishing, advertising, and science occurred in 1973 with the release of Wilson Bryan Key's book *Subliminal Seduction*, which pointed to such things as the word "sex" in a print ad, faintly spelled out by the arrangement of ice cubes in a glass of liquor. In a later book, Key reported finding a restaurant place mat that teased customers with images of an orgy hidden inside a photo of a plateful of fried clams. He claimed that these and other subliminal erotic and/or occult images that were coming in under the radar of conscious thought could only be seen clearly with an "anamorphoscope," a mirrored cylinder. Nevertheless, Key said, they made their imprint on the subconscious mind.

Some people still think Key made an important discovery. Others believe he had a bad case of sex on the brain. The consensus, though, is that he was promoting a myth. But even a myth can keep stirring belief, and get regulatory agencies into gear. After the Federal Communications Commission found itself fielding a huge public outcry, it convened hearings in 1974 that resulted in a policy statement: "Subliminal advertising is intended to be deceptive, and is contrary to the public interest."

What makes neuromarketing different from these earlier attempts to combine marketing and technology—besides the far more sophisticated and expensive equipment it requires—is that evidence coming from fMRI studies has created a fundamental shift in the scientific community. It has convinced more and more of the gatekeepers—those who hire professors and those who underwrite research grants—that the study of human emotions has now become eminently doable.

Brian Knutson, now a research star at Stanford, is a pioneer in studying emotion scientifically. He actually languished for some time on the academic job market until the concepts he was pursuing gained acceptance. He had earned his doctorate from Stanford in 1993, and he did postdoctoral studies for the next seven years while trying to land a faculty position. Suddenly, in 2000, eight invitations to interview came his way, along with four job offers.

"Oh my God," Knutson said to me in his Stanford office, pretending to be someone who had just discovered his brain imaging research, "we can *see* those regions. And Knutson is seeing stuff in those regions that is related to *incentive processing*."

Incentive processing is the aspect of brain study that will ultimately give us answers to that question, "What was I thinking?" It looks at what your brain does when it is trying to determine whether an available choice will result in something good happening—food on the table, sexual pleasure, money in the bank—or something bad. Advertisers would love to understand incentive processing, of course, and also how to use it in making their products appear more desirable.

Early in their work, Knutson and his fellow researchers realized something that casino pit bosses and lottery ticket vendors have known for a long time: Money is a really great way to get people's brains excited. Much of the work at their Stanford lab involves tracking how our brains react to spending, losing, or earning money, and even just gazing at the stuff.

Private companies are also researching this power of money to cause transformation, but they're driven by incentives that bring a whole different dynamic into play. University research is published and reviewed by academic peers. Dissenting voices are given a reasonable forum. This greatly increases the likelihood that those published findings will be solid knowledge that we can build on in the future. Private research is done in search of competitive edges, and any company with a marketplace advantage is going to keep it well hidden. Privately commissioned research is almost never published, or scrutinized by a competitor—unless that competitor has a successful corporate spy on its payroll.

The handful of firms that offer neuromarketing consultations are generally reluctant to name their clients, though they are willing to say that their rosters include big-time players. Universities usually want help in paying for their multimillion-dollar neuroscience equipment, so they often lease time in their labs to these private companies. That's our main source of evidence that extensive commercially based research is already in progress.

The Boston ad agency Arnold Worldwide provides a rare example of a small private study that got released. It recently used fMRI at Harvard's McLean Hospital to see what would happen in the brains of six men between the ages of twenty-five and thirty-four, each a whiskey drinker, when they looked at images being proposed for a 2007 campaign for Brown-Forman, their client who owns Jack Daniel's. (Other Arnold clients include McDonald's and Fidelity.)

Jack Daniel's is a bourbon whose flavorings include a strong note of charcoal. For decades the brand's advertising aimed to create backwoodsy appeal, showing older rural guys in bib overalls who looked like they grew up in a Tennessee cabin, the kind of men who knew in their bones what good bourbon was. However, because Jack Daniel's has been a perennial favorite of many rock bands—they sometimes just call it "Jack"—it is wildly popular with young people who don't mind spending their money and brain cells in pursuit of intense social moments. That demographic is eagerly courted by producers of liquor and beer, as their daily flood of TV and print ads proves, and Brown-Forman naturally wants to increase its share.

Baysie Wightman, head of a new task force at Arnold called the Human Nature Department, says that fMRI will "help give us empirical evidence of the emotion of decision-making."[2] She notes that while the subjects in the Jack Daniel's tests said that their favorite scenes were the rugged, outdoorsy ones, their brains actually showed much higher brain activation for pictures of young people having fun on spring break. So in this instance, fMRI not only showed that there was a gap between what the consumers *thought* that they thought and what they *really* thought, but also proved what images would fill that gap and score the best connection to the market.

Closing the gap between reported opinions and actual market behavior is one of the fondest dreams of the advertising and marketing communities.

In a classic instance of how relying on reported opinions can be disastrous, Chrysler Corporation surveyed car buyers shortly after World War II, when auto production resumed. It asked questions of potential buyers, aimed at understanding their criteria for choosing a brand-new ride. People reported over and over again that economy and reliability were top considerations, and Chrysler proceeded for the next few years to build its cars solidly and stolidly. Plymouths of that era, in particular, were close to bulletproof mechanically, but stylistically as exciting as a brick. Cadillacs of that post–World War II period featured a stylistic flair, a small rise in the sheet metal leading to the taillight, which suggested a fish's fin. People liked it. Living through the war years made them big fans of economy and reliability, but something about their response to the Cadillac tail fin signaled what they were really hungering for. They were ready for more fun, and more expressive design. By the mid-1950s both Ford Motor Company and General Motors, having read the trend correctly, came out with two- and three-toned paint jobs and ever-growing tail fins. Chrysler's models suddenly became the homely girls at the dance, and the company lost massive market share.

Much more is at stake in neuromarketing than cars, soda pop, and whiskey. Techniques used to test for brand loyalty are also being used to understand how political choices are made, and how political persuasion could be shaped more effectively. Recent tests have shown—and this will instantaneously make perfect sense out of every frustrating political argument you've ever been in—that numerous brain areas activate when politics are debated, but not so much the areas that do rational thinking.

Emotions run high when we decide how to cast our vote. The pleasure of feeling an emotional connection with someone on the "right" side of an issue (i.e., one's own) is a major reason why politics has always made strange bedfellows. The annoyance and fear that arise when we're feeling opposed explains why political debates can turn rancorous at warp speed.

UCLA researchers are now studying how subjects drawn from the two major political parties differ in their responses to specific campaign ads. Political party affiliation is typically inherited, passed down in the parcels of attitudes received from our parents. But genetic traits are also inherited, and it may be possible that there are structural or functional differences between people with differing political philosophies. Differences in brain structure between homosexual and heterosexual men have already been found. Perhaps one day we'll find a corollary regarding political choice.

FKF Applied Research arranges hookups between hospitals and universities who'd like to lease fMRI time and clients who want high-tech proof that their campaigns will work. It is part of a business plan the firm describes as catering to "clients who are looking to make a tremendous transformation in the way they look to advertising." Although it doesn't disclose who its clients are, FKF claims to work for "members of the Fortune 500" who "understand the importance of targeting a specific group of clients and engaging them on a visual and emotional level."

Joshua Freedman, M.D., a founder of FKF, emphasizes his firm's cost-effectiveness. "If you're going to spend $50 million on an ad campaign," he says, "wouldn't you want to know if the ad even gets out of the starting gate?" FKF sets up $3,000 sessions that allow clients to test an ad on ten subjects. That is roughly one-quarter of the cost of running a typical focus group.

In anticipation of the 2008 presidential election, FKF researchers along with several neuroscientists at UCLA pushed the implications of neuromarketing research too far and too fast. In a late-November 2007 *New York Times* op-ed piece, "This Is Your Brain on Politics," they claimed that their brain imaging study, which watched the brains of twenty swing voters, could draw conclusions about the current state of the American electorate. More specifically, they said it was possible to directly read the minds of potential voters by studying how their brains activated when they viewed presidential candidates.[3]

More than a dozen of the world's leading brain imaging experts wrote a broad rebuke of this flawed study, and of the *New York Times* for deciding to print it. They agreed that the potential use of brain imaging techniques to better understand the psychology of political decisions is exciting. But they also insisted that we can't definitively say whether a person is anxious or feeling connected simply because we've seen activity in a particular brain region. This is so because brain regions are typically engaged by many mental states. An absolute one-to-one mapping between a brain region and a mental state is not possible. For example, the FKF/UCLA article mentioned activation of a brain area that's strongly associated with emotions. But it assumed that the activations were signs of heightened anxiety levels. However, the same area is also activated by positive emotions. It takes careful experimental design to avoid misinterpreting brain imaging studies. And, the scientists added, "as with any scientific data, the peer review process is critical to understanding whether the data are sound or based on faulty methodology."

Knutson has a strong appreciation for careful experimental design, heading up a Stanford research group called SPAN, the Symbiotic Project on Affective Neuroscience. ("Affective" refers to emotions.) SPAN was conceived to build greater understanding of the physical basis of emotions and emotional expression in the brain.

On the warm August afternoon in 2007 when I entered the Palo Alto campus office building where Knutson works, I walked past two statues in the entryway: Alexander von Humboldt and Louis Agassiz—two of the greatest interdisciplinary minds in the history of science. Humboldt's quantitative work on botanical geography was fundamental to the field of biogeography. Agassiz greatly expanded ichthyology, the study of fish, though he is best remembered as the first person to propose scientifically that the earth had been subject to a past ice age.

Those two pantheon scientists would find brain imaging extremely exciting, for the same reasons that it excites Knutson and so many other contemporary scientists. So many different fields are meshing, creating new hybrids, and so many impressive discoveries are emerging as a result.

"I basically had a normal Beaver Cleaver upbringing," Knutson says of his young years in Kansas City, Kansas. After a reflective moment he adds, "But, having said that, I never felt that I fit in where I was being raised. Many teenagers feel this way. But I felt it pretty strongly. I had a lot of friends but never really belonged to any one group."

Although I'm sure his memories are true, you would never guess it from his at-ease demeanor, and the openness with which Knutson gave me over two hours of his jam-packed schedule, offering illuminating comments on everything from his latest research to the tiny pair of Hindu deities (Shiva and Ganesh) who overlook the office from his bookshelf.

While studying at Trinity University, a small liberal arts college in San Antonio, Texas, Knutson searched for answers to his perceived absence of emotional connection. He gravitated to the field of psychology and did experimental work with Dan Wagner, who would later join Harvard's faculty. Knutson also spent time in Nepal, studying Buddhist communities. "When I learned about other religions, it just blew my mind," Knutson says. "Especially Buddhism, because of its worldview, was really interesting and possibly useful."

Wagner was conducting studies about how to deal with unwanted thoughts mentally and emotionally, a process called thought suppression. "What we kept finding," Knutson says, "is that you could predict huge individual differences depending on a subject's overall mental state." Specifically, people with depression had an extremely hard time fulfilling an instruction like "Suppress the thought of a white bear."

"So," Knutson says, "I started to think of how important emotion is." This idea connected powerfully with his religious studies. "The center of the Buddhists' psychological universe is attachment," he says. "When Buddhists use that word, what they mean is how you *react* to things. A lot of Buddhist practices are about changing how you react, developing mental and emotional habits that are better for you in the long run, more peaceful and constructive."

Knutson's interdisciplinary background—he took degrees in both experimental psychology and comparative religion—helped him win a fellowship at Stanford, where he earned a psychology doctorate in 1993. "The data kept whacking me over the head and suggesting that emotion was important and I should be studying it," he says. "But the problem was that it was always very difficult to measure. I realized that if I really want to understand emotion, I've got to understand the brain."

To feed his curiosity, Knutson enrolled in med school classes. One of his professors recommended that he work with Jaak Panksepp, one of the very few people then studying how emotions work in the brain.

Panksepp was teaching at Bowling Green State University in Bowling Green, Ohio, a few miles south of the shores of Lake Erie. He was born in Estonia but grew up in the United States and began in the 1960s to study the brains of rats. In 1998 he wrote the pioneering book *Affective Neuroscience*.

"It was a radical departure for me at that point," Knutson recalls. "But it was great. From a career perspective, it was a turning point. We gave drugs to rats and looked at their brains and looked at their behavior."

Completely by accident, Knutson discovered something amazing during those trials. "I stumbled upon the fact that the rats were making some vocalizations when they played," he says. "Rats play, just like humans. But the point is that these vocalizations I accidentally discovered turned out to be centrally related to emotion.

"Panksepp had always argued that rats feel emotion, but it's very hard to know just what a rat is feeling. Well, it turns out they make ultrasonic vocalizations related to emotion, but in a frequency band that people hadn't been paying attention to."

The rats made their vocalizations when they thought something good was going to happen, like food or some other reward. In other words, they were doing incentive processing. Panksepp and Knutson injected dopamine—a hormone and neurotransmitter that stimulates the brain's pleasure system—in a specific area of the test rats' brains. Every time they did, the reward-related vocalizations came pouring out, happy squeals at extremely high frequencies.

Knutson also decided to run an experiment on himself while at Bowling Green. He was collaborating with psychiatrists from the University of California, San Francisco, about selective serotonin reuptake inhibitors (SSRIs), drugs of the Prozac and Zoloft family, which are frequently prescribed to treat depression. These drugs work by rebalancing brain chemistry, and they can be effective for many other ailments besides depression. The researchers wondered: "What might happen if we give normal people SSRIs? Would that change something about their emotional life?"

"And," Knutson says, "we actually found out that it did. Since I was giving that medicine to other people," he adds, "I thought I should take it myself." He did. "One day I was walking down the street, and it was frosty cold. The wind comes off Lake Erie in the winter and really hammers Bowling Green; that was one of the reasons I didn't like living there. When I saw two people walking past me, all of a sudden I had this thought: 'Oh, they look happy. It'd be kind of nice to live here the rest of my life.'

"Having that thought really surprised me. It had never entered my head before then. But it was really consistent with the results of our study. I was already a fairly happy person, but some edge of existential angst had gone. I had stopped struggling with the feeling of, 'There must be something more to life.' "

The importance of the rat/dopamine discovery, along with a decision that he needed to move along if he wanted to further his career, helped Knutson land a postdoctoral position at the National Institutes of Health. One day in 1997, an adviser there mentioned to him that some time was available for use of the lab's new brain imaging equipment.

"That's how I got into fMRI, and I just went nuts," Knutson says. "Immediately I wanted to *see* those deep areas in the brain that you could stimulate to get these vocalizations. It wasn't easy back then, but it paid off in the long run." He found out that signals from those areas in the inner, more primitive parts of the brain tend to increase when people think good things are going to happen, which was a finding he expected. But he also learned that the signals *predict* interesting things because they would occur when subjects were about to take a financial risk or were about to buy something.

"These are pretty sophisticated abstract behaviors," Knutson says. "Science didn't traditionally associate these kinds of behaviors with deep limbic structures."

In other words, activity in the oldest parts of our brains can stimulate activity in the newly evolved areas. Without knowing it, our most sophisticated and evolved brain regions take a lot of their orders from brain parts that haven't changed much since they were the main central processing units for the tree-dwelling cousins in all of our family trees.

Not long before I visited his office, Knutson published a study on predicting shopping behavior, a job that had consumed his entire yearlong sabbatical. "Shopping is a decision that all of us make all the time," he comments. "And yet there was little out there on the subject that bridged neuroscience and economics. Economics has very sophisticated, sometimes beautiful theories about things. Neuroscience has very basic building blocks. So I was trying to bridge these two." The result was the first study to predict purchasing on a trial-by-trial basis.

Scientific American reported on the study, which involved twenty-six subjects taking part in something called SHOP, an acronym for "save holdings or purchase."[4] While lying in the fMRI, each subject looked at a screen on which a typical product would appear, followed after a four-second interval by a purchase price. After four more seconds elapsed, two small boxes would appear. On one side of the product picture, the subject could choose a box marked YES. On the other was a box that said NO. When those boxes appeared, the subject, whose brain was being scanned, had to choose whether or not to make the buy. The boxes appeared on opposite sides at random. In the majority of trials, the buying and rejecting were imaginary, but in two trials the subjects had a $20 grubstake and could actually claim their purchases.

The reason Knutson invested what could have been a year of free time in designing the experiment and carrying it out was that he wanted to find out if he could predict individual purchasing behavior, the holy grail of every marketer.

The scans showed that a certain brain region switched on when the product shots first came up, a midbrain area that seems to be part of our mental reward centers. When the price was revealed, an evolutionarily newer brain area spoke up, a structure that gets involved with weighing decisions. Another area that activated, in a different region, seems to be a device for sorting out negative stimuli. It activated most strongly when subjects chose to click on NO.

In short, by studying which regions were activated, the research team was able to predict whether the participants would decide to purchase each item.

The implications for marketers could be wide-ranging, including the possibility of establishing fMRI-based evaluation of advertisements and prices. For example, a recent study by Knutson and Cal Tech colleague Antonio Rangel showed that people perceive that they will enjoy a wine more when it has a higher price. "We find the more expensive the wine, the more activity we find in the medial orbitofrontal cortex of the brain," said Rangel, an associate professor of economics. "I can change the activity in the part of the brain that encodes for subjective pleasantness by changing the price at which you think the product is sold, without changing the product," he added.[5] (It should be pointed out that some neuroscientists have disagreed with these conclusions.)

Knutson is not concerned with accelerating consumption, but rather with understanding decision making and ways to help people make more intelligent decisions.

For example, credit card purchasing often leads to judgmental errors. In a credit purchase we don't connect as completely to the reality of how the expenditure reduces our financial reserves. It's kind of abstract, and thereby leads to extra spending, and a dearth of savings. A healthy savings rate is, however much we may resist it, a key to financial independence—on both the personal *and* the national levels. By knowing the brain activations involved, Knutson hopes, we may find more effective ways to keep buying from leading to remorse. While you might try chanting, "C'mon, insula, do your thing!" or simply telling your medial prefrontal cortex to be more assertive, it will probably be more effective to take a brain training course of the near future and learn how to quiet the brain area that's bidding for instant gratification.

Whatever the prevention techniques may turn out to be, neuroscience has not only validated common sense about spending behavior, but shown exactly what brain areas can cause us to be "of two minds" about financial decisions, and to understand that impulse purchases are dictated by a relatively primitive brain region that is very good at getting itself heard.

The common thread across Knutson's career is a set of deep philosophical questions in search of scientific answers: What drives people to do the things they do? If behavior is rooted in the structure and chemistry of the brain, how do you deal with it? Do you just do what your nervous system tells you to do, or can you do something different? How do you change your brain processes for the better? How does the machinery work?

To illustrate his point, Knutson encircles his cranium with both hands.

"This is the mechanism," he says. "I don't know what all this research will lead to, specifically, but all I can say is that understanding the mechanism will lead to better interventions." And these interventions will radically improve lives throughout the world, not just in terms of financial behavior but across a wide spectrum of concerns, including mental illness.

Earlier, while weighing the possible long-term impact of advancing brain science, Knutson had said, "Everything's at a very young stage. In a historical context, we've had fMRI for maybe a little over ten years, and this decision-making stuff, about four or five years. So the growth is remarkable. And, of course, the hype is going to supersede the growth. People are going to get unrealistic expectations about things. But I wouldn't be betting my time and money and career on this if I didn't think it holds some promise."

Read Montague of Baylor is another intensely busy researcher. Like Knutson, he has written some of the most frequently cited papers in their field, and his work has already had a huge impact on both academics and businesspeople. Both groups really started paying attention after Montague published an amazing study in 2004 about corporate warfare, specifically the age-old dominance struggle between Coke and Pepsi.

Before he committed to a science career Montague was a four-sport athlete in high school, went through college on a track scholarship, and competed successfully enough as a decathlete to ponder an Olympic tryout. He understands that competition in sports gives high rewards for quick responses, and the reasons some coaches tell their players, "Don't stop to think: You might not get started again!" A sharp, maybe even killer, instinct for where the ball is about to go, or what weakness an opponent will reveal, can crack a ball game wide open in a heartbeat. Pausing to think about what to do next can put it in the tank.

Since life itself is a competition known as survival of the fittest, our brains want us to be effective warrior-athletes, for the sake of our self-preservation. They want us to make quick decisions that will get us out of harm's way, and into situations that keep us alive and our DNA in circulation.

Montague's Coke-versus-Pepsi study was aimed at understanding in detail how our rapid-response brain mechanisms work, and how we might counteract the ways in which they sometimes work against us. "I am going after will," he says. "I'm going after willful choice. It just happens to have a lot of practical implications."

Lots of marketers jumped the gun, trying to draw highly specific conclusions based on what emerged from the study. It wasn't intended to demonstrate that one brand was better, another one worse. Montague wanted to know something basic about how our brains make choices. Evolution gave our brains the power of quick, instinctive response for some very good reasons. Now we need to understand how that power works in modern life.

"When you eat a berry," he explains, "if the berry is shiny and red and looks ripe, and the inside is good, then you'll experience the berry in a certain way. But if you change how the berry looks on the outside, your experience of the berry will be different—even if the insides are exactly the same.

"There is a reason for that," he adds. "A bush is *marketing* something. It has to attract birds and other creatures to eat its berries, so the seeds will get carried off to new ground. Well, it can't print an advertisement that says 'You should eat this!' Instead, in an evolutionary sense, the bush learns how to tickle the bird's nervous system in a way that makes the bird *want* to eat the berry. It wraps the berry in a carbohydrate jacket that the bird would like, and finishes it off in a color that is, basically, a brand. As soon as that bush has made the berry ready to be eaten, it turns the berry red."

In other words, marketing isn't an invention of capitalism, or of any other economic system. Marketing comes straight out of nature.

Like birds, bees, camels, and practically every kind of organism, humans carry around circuits that alert us to hidden value. The color of the berry, like the design of the packaging that turns up in our supermarkets, is a proxy. It represents satisfaction. It says, "This is a great choice for taking care of yourself, of your need to consume food."

In the deeper and more primitive areas of the brain, signals like the shape and redness of a berry come in and take command. Then they take command of your behavior. Just as the bird clutches a ripe berry in its beak, we put an attractively packaged item in our shopping basket, or in the virtual shopping cart of a commercial Web site. Or we place an online auction bid.

Even the currency and the credit cards we use are proxies. They represent our ability to possess the many things that are out there, trying to enchant our minds into buying them. The seduction goes even further. Credit cards are a proxy for money. They act as an "enabler." Any time we pay for something by credit, an area of our brains that registers the anticipation of pain activates only about half as much as it does when we use cash.

When he designed the Coke-versus-Pepsi study, Montague wanted to resolve a question that had been bugging him. "How do these edicts, these directives, get inside your nervous system? They insinuate themselves, and they govern your behavior. They must be taking over circuits, and commandeering knowledge networks."

Montague's daughter was at that time between her junior and senior years of high school and about to do summertime lab work for her dad. Most of the experiments he designs involve computational models built around complex mathematics, engineering, and physics. He wanted to do something fairly uncomplicated instead, so his daughter could take part without needing to tackle such advanced topics.

Chemically speaking, Pepsi-Cola and Coca-Cola are about as different as Tweedledee and Tweedledum. And yet many people claim to love one and dislike the other. Montague wondered why. He and his associates rigged plastic tubes so they could squirt sample tastes of the soft drinks into the mouths of subjects who were having their brains scanned. In some experiments the brand names were identified. In others they were kept secret.

In the first go-round, researchers simply watched the evidence posted by people's brains when either of the two soft drinks landed on taste buds. This set of responses—called blind responses because the subjects didn't know which brand they tasted—ran close to fifty-fifty. Parts of the brain that respond to rewards were activated immediately by each of the two brands, and in nearly identical measure.

Then the researchers repeated the procedure, but with one crucial difference. This time they told subjects which brand they would be getting *before* the tastes arrived.

Amazingly, brain responses in 75 percent of the subjects showed a preference for Coke.

Since actual tasting resulted in essentially equal numbers, why would foreknowledge of the brand literally change someone's mind?

As Montague explains, "People are not buying the content of the can. They are basically buying the brand, or the *feeling* that the brand gives them."

In other words, knowing what brand was trickling over their taste buds added some information. Because their brains found that information meaningful, the subjects experienced a symphony of reward activation. Many of the brain areas that got stimulated were ones that have a lot to do with memory.

The study did not show that one brand really tasted better than the other. It simply showed that Coca-Cola's marketing campaigns have had more success than Pepsi's. Coke has created a stronger proxy effect. It has done a better job of linking its product with a belief that the product will give you a satisfying experience.

Coke versus Pepsi is a marketing battle that has lasted more than a century, since the two drinks were invented, separately, in pharmacies located down south. Atlanta spawned Coke, and New Bern, North Carolina, gave us Pepsi.

In 1940, Pepsi was first to air a jingle nationwide by radio. Coke's extensive advertising over the years standardized the popular image of Santa Claus as a jovial guy who happens always to dress in Coke's signature color, red. Catchphrase fusillades have come and gone, telling us that Pepsi is "the Choice of a New Generation" or "the right one, baby," while Coke has advertised an ability to "teach the world to sing / in perfect harmony."

Coke's greater success might come from something as basic as the color chosen for its packaging. Dr. John Stith Pemberton, the pharmacist who invented Coca-Cola, happened to choose a hue of red. It might evoke thoughts of berries, apples, or plums, or of sexual excitement. There are powerful reasons why most lipstick is vividly red, the same color that appears when we blush or are erotically stirred. It's the color of blood coming to the surface of human skin, signifying arousal.

Creating a proxy effect is obviously powerful. It's the basis of a natural survival strategy called mimicry. In 1852, a British scientist named Henry Walter Bates was studying butterflies in a Brazilian forest. He found two that looked very much alike, but after examining them closely he realized that they weren't even related.

One of the butterflies happened to be deadly poison to any bird that ate it; the other was not poisonous. That second butterfly, though edible, had evolved a proxy effect. By mimicking the looks of the poisonous butterfly, it sent birds a deceptive message. Something like: "Remember me? I'm the bad boy that can kill you."

It isn't surprising that humans would adopt mimicry. Our nervous systems evolved from the same types of nervous systems as those of birds and other animals. We have, by hook or crook, discovered ways to do the same kinds of tricks. For example, car and truck dealers like to photograph whatever they're selling alongside an invitingly curved young woman in a bikini, leaning against a body panel with her smile flashed onto high beam. Why is she there? Well, her sexuality obviously helps close sales. Some potential buyers, in deep and primitive structures of their brains, see the car or truck as a proxy for reproductive success. Some aggregation of cells that evolved in the days when the big competition wasn't Coke versus Pepsi, but rather Neanderthal versus Cro-Magnon, calls out: "Want woman! If get Corvette, get woman, *too*!"

We may never know exactly why Pemberton chose red for Coke. But in the near future, such choices will probably be made after a review of brain scanning data. Advertisers pulled two crucial lessons from Montague's study.

The first is that if you work long and hard and smart enough to build a brand's identity, you can wire that brand into the brain's happiness circuits. When you do, opponents will be blocked out of that desirable neural real estate.

The second is that neurotechnology can produce feedback about how good a job you're doing at creating a proxy effect. Then you will know whether or not you need to fine-tune your marketing efforts.

For the last four years, Montague and his staff have run experiments to shed light on a simple but profound question: After someone has done you a favor, like, for example, sponsoring the study you're about to participate in, can you still have independent judgment? Can there be any hope that the imaginary researcher from the start of this chapter, studying Hilda's Tastee Soups, could override bias and tell you whose cream of tomato was genuinely the finest?

Montague's experiments were handled in such a crafty (though totally ethical) way that the subjects being tested didn't have a clue as to what kinds of facts the researchers were really after. First, the designers of the experiments assembled images of Western art, from paintings widely acknowledged to be among the greatest. These included both representational and abstract images. Then the scientists invented some fictitious companies, giving them professional-looking logos and Web sites, even active bank accounts, trying thoroughly to make them seem real. Subjects would lie in a scanner and see the art pieces come up on a screen for five seconds, one after the other, in random order, and rate each on a nine-point scale, from minus four to plus four, depending on how much they liked each one.

When the subjects got positioned in the scanners, the first screen they saw announced that a certain company was sponsoring their session, and would pay them $30 for participating. They didn't know the company was imaginary. Its logo was shown alongside the words of the announcement, just as one would expect. From time to time through the experiment, the logo of the sponsoring company would show up again, and so would the logo of another company, one that didn't "sponsor" the subject's test. Subjects were told to rate the pieces of art, just as in the prior experiment. No additional instructions were given.

The researchers wondered if subjects would rate artworks more favorably if they showed up together with the logo of their "sponsor." And in fact, that's what happened. Being visually associated with the sponsor resulted in more positive valuations. Doing favors, therefore, pays off for the giver. Sponsorship and impartiality don't actually go together. People remember what side their bread is buttered on.

The implications are obvious. "Since I'm at a medical school," Montague asks, "what happens when Big Pharma pays for a science study? What happens when Big Pharma gives me any money at all? Can I ever read a paper sponsored by them objectively? The answer is no. A person can't just think his way out of being influenced. So the big issue on the table is, how do we keep scientific progress happening? Can we actually keep commercialism totally out of science? Well, there is no way to do that. But to pretend like money doesn't completely bias you is naive, if not totally dishonest."

The application of this research, Montague suggests, is to be wise about how incentive structures are set up and to utilize scientific methods like "blinding" the experimenter to eliminate bias whenever possible. The system should advance knowledge itself, and not corporate or government agendas. In the long run, biased experimentation just leads to useless data.

"There are key places where your will is sort of bled from you," Montague says. "We don't have complete free choice. We don't have a totally independent will."

Montague's track record helped convince Baylor to raise funds for three more 3T MRI scanners, the most powerful that are currently available. 3T means 3 teslas, a measurement of scanning power. Having a five-unit complex of scanners is going to allow Montague to set up a social imaging cluster, a resource previously unheard of, especially in the currently tight state of funding for science.

Montague wants to learn how our brain processes signals and information in the context of a group. He believes that a large constellation of mental illnesses, all the way from mild anxiety up to severe depression, will be better understood by scanning brains that are interacting. If he's right, the upcoming studies will have tremendous applications in the medical community, and for society in general. "We will have the place," he states, "and the equipment, to pursue that kind of knowledge. Hopefully, that's how we'll pay for this equipment. If not, I'll probably get fired. But before that . . ." He lets his joke float in the air.

Before he gets fired (something extremely unlikely to happen) Montague will take a run at a provocative batch of questions, such as: If you're in a group, how do you respond to the presence of an authority figure, or the presence of people of other races? Who in the group becomes the dominant one, and who becomes less dominant? How do we respond to being labeled by others? What makes us *want* to put a label on someone else? How much is our will swayed by our need to feel like we're a part of something?

Needless to say, marketers will be just as excited as professionals in medicine, government, and other fields to get the answers to fundamental questions like these about the brain.

Social pressure can sometimes make us do or say something we didn't expect to. We may wonder: "Why did I do that? I hate it when I'm doing that! Why did I say that? Why did it come out of my mouth? Why did I speak up?" It's a strange sensation, but probably universal.

The answer, according to Montague, is that your behavior can be dictated by mechanisms inside your brain, independent of your will, that step in and take over. The more we can understand exactly how that happens, the more we can counter the effects of those mechanisms and be true to ourselves.

"I'm going after what affects your capacity to make an independent choice in the context of a group," Montague says. "The only way to do it is to set groups into interaction, and then study what happens inside the interacting brains.

"Everybody can't be the chief," he adds. "When there is a big group, people have to stratify. Somebody's got to follow. How does the group implicitly decide that? What happens when somebody takes over? Better yet, how does someone succeed in selling some whacked-out idea to the group? 'Hey, why don't we go deny every biological instinct we have for survival, and kill ourselves by blowing up the American ship?' "

Montague calls terrorists "idea entrepreneurs" because, in a sense, they are the world's most incredible salespeople. They convince people to embrace their own death.

The subject of terrorists raises some profound questions. Neuroscientists are scientifically establishing some major points about how persuasion is made to happen. As our knowledge of the brain advances, will we see the technology, and the advanced knowledge it produces, used in greed-driven, or hate-driven and murderous, ways?

It's a simple question, but direly important. It hangs over every scientist now contributing to the field's progress. Ultimately, is neurotechnology going to fulfill great promise or expose us to extreme peril? Or will it deliver both things, but in a fluctuating blend?

"I think," Montague says about the transformation that's happening to our knowledge base, "the analogy is with fertilization techniques. If you look at books from a hundred years ago, you will see drawings of sperm cells with little babies curled up in the end. Now we know about DNA and reproduction, and we can literally grow whole organisms in about three weeks from an embryonic stem cell. We can engineer genes that do all sorts of things, and on and on and on. This has absolutely changed our view of ourselves. A lot of things that were left to 'God,' or to the mysteries of the Great Out There, are now in our control. And it's kind of frightening. We don't always know what to do, what would be moral or immoral. I think the problems in genetics research are small compared to the problems that will happen as neurotechnology matures. And it's going to mature faster than people imagine.

"In fact, cognitive neuroscience has the biggest 'peril' stamp on it ever. If it really works, then it's like nuclear energy. It could do great things, or evil things, or both. Let's be straight about it.

"But that's what we are paying for—to move neuroscience ahead."

And moving ahead it is. Each year more powerful advances in brain scanning technology emerge, and those advances get into the hands of more and more researchers. With Time's Telescope as our guide, we can see that fundamentally new neurosensing technology will emerge over the next decade driven by the extraordinary economic value inherent in understanding the human mind. Upcoming generations of scanners will make today's machines look like the old computers driven by vacuum tube technology, compared to the infinitely faster microprocessors of today. Where will these new capacities put the field of marketing and decision sciences twenty to thirty years from now?

Clearly we will uncover some core knowledge about how our brains make decisions. Moreover, noninvasive neurotechnology for uncovering our underlying emotional states will become commonplace. For example, one way that neurosensing technology will soon make its way into the lives of hundreds of millions of us is through the expansion of lightweight brain sensing equipment attached to near-future video gaming systems.

The $28 billion video game industry has already latched on to neurosensing technologies and tied them to futuristic video games. Silicon Valley's Emotiv Systems has developed a new interface for human computer interaction. Project Epoc is a highly stylized EEG system that connects wirelessly with all game platforms from consoles to personal computers. NeuroSky is another developer of sexy brain sensorware that collects brain wave signals, eye movements, and other biosignals, which are captured and amplified, via their patented dry-active sensor technology. While NeuroSky's headset has one electrode, Emotiv Systems has developed a gel-free headset with eighteen sensors. Besides monitoring basic changes in mood and focus, Emotiv's bulkier headset claims to detect brain waves indicating smiles, blinks, laughter, even conscious thoughts and unconscious emotions. Players can kick or punch their video game opponent—without a joystick or mouse.

In years to come gamers will bid good-bye to their joysticks, controlling play with their thoughts, as these neurotainment technologies pump up the extremely cool factor of cutting-edge game experiences. Marketers will seek to tap the information generated from the systems when used in a massively multiplayer Internet-based gaming world to test brain-based reactions to new product messaging among their next generation of consumers, perhaps even paying the players for access to their microreactions.

Future generations of the Nielsen rating system and other entertainment rating systems will also incorporate noninvasive emotion-sensing technologies without the need to wear any gear, just the addition of a tiny video camera. Leveraging the brilliant facial expression research of Paul Ekman at the University of San Francisco, the Department of Homeland Security is now developing a system that can uncover hidden emotions, such as those caused by lying, by capturing fleeting "microexpressions" that are the result of the millisecond movements of forty-three muscles in the face. Ekman's research is already used by computer animators to create realistic facial animations and by police officers interviewing suspects. It will soon enter many more of our lives in more powerful ways. For example, this technology will further accelerate the mass customization of advertisements, tailored to the emotional state of the viewers. We may call this new field neurotizing.

To counter these more sophisticated techniques, neuromarketing alert systems will emerge to identify when and where these crafty, subtle techniques are being employed. Some governments may even go so far as to screen and label neurotizing with specific warnings about its intimately invasive nature, to protect undereducated, unaware communities from the rapid expansion of these technologies.

Within education, the science of decision-making and neuropersuasion technologies will spark the creation of incredible social value. We will be able to reevaluate and redevelop how we educate our children in a world of exponentially expanding knowledge. Teachers and parents, with the aid of unobtrusive, neurofeedback equipment, will know when a student's mind is at its most receptive. Corporations will also use this technology because lifelong learning will be a hallmark of our hypercompetitive world.

Today we're fixated on the SAT (Scholastic Aptitude Test) for measuring academic potential. In the future we'll likely have the BAT (brain aptitude test). It will test not only a student's level of knowledge but also her innate reactions to the presentation of new knowledge and new circumstances, based upon her continuously changing neurobiology. Many employers will want to know how well individuals will perform under novel stress. Because these teaching and testing technologies could also be used for nefarious purposes, and will be widely dispersed, keeping them out of the wrong hands will be a tremendous challenge.

On the bright side, our newfound knowledge will lead to whole new sets of impulse-control techniques and technologies to help us manage our illogical spending patterns, proneness to compulsive shopping, and even irrational stock-trading tendencies. These will come in the form of neurosoftware systems for retraining ill brains, neurostimulation devices that sense oncoming impulses and lightly jolt the brain into a less compulsive state, and drugs to calm the entire mind down.

Now that you have a vision of how all these new technologies will emerge in marketing, you may already be wondering: How will neurotechnology impact the evolution of other parts of the business world, more specifically, the global financial markets?

FOUR

FINANCE WITH FEELINGS

It is the emotion which drives the intelligence forward in spite of obstacles.

—Henri Bergson

Lack of money is the root of all evil. —George Bernard Shaw

In the summer of 1995 Richard Peterson was twenty-two years old and a brand-new University of Texas graduate. He sat on his family's front porch in his hometown of Lubbock, thinking over his ambition to go through medical school and become a psychiatrist.

All that schooling was going to be very expensive. But he had a computer-driven idea for beating the stock market, and he was fairly confident that it would help him foot the bills. He'd written software that was supposed to make stock market investing as purely rational as possible. That idea made good sense according to the classic theory that ruled economic thinking a decade ago, before brain imaging proved that it was solidly wrong. Peterson couldn't have known it then, but by pursuing his idea he was about to become one of the frontier scouts of a movement that will ultimately hit society like a worldwide gold rush and change our thinking about money forever.

Conventional economic theory has long maintained that our financial lives are ruled by *Homo economicus*, a completely rational character who lives inside each of us and who makes all our financial decisions. And yet we've all made financial decisions we thought were rational at the time, but which turned out wrong and left us wondering why.

Now that brain imaging allows us to peek at the complicated dance that goes on between emotions and judgment, we're moving toward more realistic new theories about how our financial brains work, and also toward evolving some practical methods for spinning profits out of all this newly discovered knowledge. Researchers have evolved some new fields of study. The financial realities of the coming neurosociety will be determined by the facts being learned in those fields.

Neuroeconomics, and its close cousin neurofinance, are still in a developmental stage. But it's a developmental stage that is being closely watched by people who are excited about finding new competitive advantages. Neuroeconomics research relies on real-time pictures of brain activity, such as fMRI. Neurofinance uses brain imaging and neuroscientific principles to understand how to improve financial trading performance.

To date, just a few financial traders have tried to apply the knowledge coming out of research labs. There are also still plenty of financial professionals in the world, and economics teachers, too, who have never heard of neuroeconomics or neurofinance. We can expect both of those things to change rapidly.

The development of brain imaging made these two new fields possible. Think of them as vigorous crossbreeds. They mate the sober discipline of economics with the reality-juggling methods of psychology lab experiments. Both fields have repeatedly proven something every one of us needs to understand about handling money: Whenever we believe we're *rational* about financial choices, we're gripped by a dangerously *irrational* belief. Emotions constantly distort our attempts to be smart with money.

But those same emotions, if understood and harnessed, can revamp our style and help us make far more profitable choices.

Neuroeconomists look at economic behavior, just as traditional economists do. But their greatest preoccupation is watching the brain's circuitry morph off and on, reacting to the moral, ethical, and financial decisions designed into their experiments. They probe both the how and the why of the brain's reaction to money and choices, looking for predictable patterns. Their research takes them into some of the deepest and most secret-laden levels of the human psyche.

These new fields are a key indicator of the truth that a neurosociety is really on its way, based on evidence we can see in Time's Telescope.

We may think of people in the financial professions as being incredibly conservative and some of the last folks likely to embrace anything new. But that idea is as wrong as the one about *Homo economicus*, the guy who is totally rational about money decisions. Believe it or not, people in the financial fields push themselves up to the front of the line for technological breakthroughs. They are some of the earliest to seek ways to leverage inventions into advantages, and because of their determination to adopt those inventions, new ideas and methods start spreading quickly through an absolutely necessary aspect of human society, the economy.

Especially in fairly recent history, financial professionals have been fast to jump aboard new technologies. They live in an extremely competitive world, and they're always hungry to be first in finding a competitive edge. In the earliest days of the Industrial Revolution, canals were dug across the landscape from city to city to ease transportation of factory goods. Banks immediately began using those newly built canals for timely, cost-efficient movement of gold ingots and trade news. A few decades later, railroads and telegraphs made it possible for people to shunt goods and information even more rapidly. Suddenly, local banks were able to concentrate their financial power by evolving into national networks. The banks that evolved became giants. When calculators, typewriters, telephones, and electricity came onto the scene, we suddenly had international stock markets. This created more desire for the competitive advantages that can result from rapid movement of timely information. Then came microprocessors, with their ability to move information at hyperspeed and to crunch numbers just as quickly. These capabilities brought forth several brand-new investment possibilities, including mutual funds, hedge funds, and derivatives. With history as our guide, we can see that financial professionals will also be early adopters of emerging neurotechnology.

Peterson's student-era attempts to harness technology for financial gain led him into the birthplace of neuroeconomics. He had written a software program designed to predict movement in the futures market for his senior project for an electrical engineering degree. It was meant to predict which way the entire beast was going to turn, either up or down. It would guide him in placing quick investments on the anticipated rise or drop of each coming day's market, much like choosing either red or black at a roulette wheel. The software also dovetailed with the cognitive philosophy degree Peterson earned simultaneously. Most of us might think electrical engineering and cognitive philosophy are utterly different realms, but in Peterson's lively mind they converged as a beautiful blend of engineering and art.

Peterson had actually begun playing the stock market at age twelve, and he had done quite well. His father, also named Richard, a former Federal Reserve economist who was a Texas Tech professor, had started him off in the investment world. Peterson senior also joined in the computer-driven plan his son invented. He contributed some ideas for the software, put some cash into the investment pool, and also brought a colleague in as an co-investor. They named their enterprise Intelligent Investments.

Peterson the Younger, meanwhile, sold all the stocks he had accumulated since age twelve, generating nearly $60,000. He put $25,000 into Intelligent Investments, which began with total funds of $50,000. The remaining $35,000 of his stock earnings got set aside for another investment idea that intrigued him. It was based on intuition, but it needed further development.

Intelligent Investments ran by this plan: At 3 p.m. local time the stock market closed. The futures market closed fifteen minutes afterward. Peterson phoned his broker during the brief gap to find out closing cash prices. He ran the numbers through his software, and out came a prediction. With the last of the fifteen-minute gap ticking away, he would hurriedly trade futures based on whether the computer thought that tomorrow's market direction would be up or down, red or black.

It worked. Over the first few months it proved correct 58 percent of the time and yielded an 8 percent return. Not bad at all. However, 1995 was also the start of a long, truly incredible bull market. Peterson would've made a lot more by investing in good stocks and holding on. But he saw his plan as Version 1.0 of a perpetual-motion moneymaking machine. He even thought he might postpone med school to explore its potential further, especially after a sweet string of twenty-four days in which the system was 100 percent correct.

Meanwhile, the intuition-based idea kept teasing Peterson's imagination. In a side experiment, with no money attached, he casually recorded his daily gut-feeling market predictions. His intuition-based model was based on four educated guesses. It worked like this: At the end of trading each day, Peterson recorded his own emotional state and his gut-level market forecast. He also noted how confident he felt about that estimate, within a five-point range. Finally, he compiled something he called "the mood of the market," a totally seat-of-the-pants, impression-based concept combining recent financial reports from newspapers, television, and other random inputs.

After three months of computer investments, and also writing out his gut predictions, Peterson compared the numbers. The intuitive method had scored 70 percent accuracy, better overall than the computer software. Then, in a single week during December 1995, the computer-run system hit a string of bad days and lost close to $10,000. Fearing a wipeout, the three partners of Intelligent Investments decided that closing their enterprise was now the intelligent thing to do. Peterson still had his original $25,000, plus $12,000 in profits, as well as the $35,000 he'd previously set aside. He was ready, and stoked, to test the real-life accuracy of his intuitive model.

He began in January 1996. Great results tumbled in immediately. Out on a first date in February, he heard himself bragging about the profits. One week later his new girlfriend innocently asked, "How are you doing now?" He did a quick mental review and experienced a cold shot of reality. "Uh, not as well," he admitted. The math forced him to acknowledge that he'd just lost almost $6,000 in a week. A few days later he was eyeball to eyeball with this curious fact: The intuitive model had started running *wrong* 70 percent of the time.

He went back to the computer-run plan. The returns were low but more consistent. But even with his intuitive plan shelved, Peterson still kept recording his from-the-gut predictions. By the end of February they had bounced back to 70 percent correct. This exciting return to form lasted throughout March, April, and May. So in June 1996, when he was entering med school, Peterson began round two of backing the intuitive model with money.

Once again, his intuitive predictions started out great and then rapidly fishtailed back to 70 percent wrong. He shelved it again but still recorded the data and wondered what was changing his results.

Then, as December 1996 arrived, the intuitive model began to generate powerful indicators that a boom was coming. Peterson gambled, borrowing money to purchase futures. On December 5, the chairman of the Federal Reserve, Alan Greenspan, gave his famous "irrational exuberance" speech, warning investors that the stock market was overheated and about to go into the tank.

"I was studying for my finals," Peterson says, "when I got a call from my broker. Prices were plunging. The money I'd borrowed to invest was being recalled." Peterson closed his position out. The very next day, though, giving the strongest signal it had *ever* produced, his intuitive system predicted that the market was going to rebound. But he reflected, rationally, "I just lost five grand while trying to study for finals. This is crazy!"

As stock market veterans know very well, a tremendous rally arrived soon after Greenspan's words of doom. Peterson realized that if he had stuck to his intuitive guns, he'd have collected enormous profits. That missed opportunity was one of his early cues about the collective power of marketplace emotion. Greenspan's speech had caused fear in people who were exuberant. A big number of them sold off their stocks, driving down prices. Wiser investors waited a couple of beats, then started to pick up the bargains.

Peterson wasn't sure what to do with that insight, though, and decided that he couldn't afford to keep playing. But even though he stayed out of the market, he kept on collecting data for his intuitive system. And once again his non-funded forecasts began to show 70 percent accuracy.

Meanwhile, all of Peterson's investment-savvy college friends were happily making thousands on Internet stocks. He was missing out on the incredible tech-fueled boom. But he still thought that mathematics and computers would ultimately be the answer, if he could figure out how to handle that emotional component.

By then he had three years' worth of data to analyze. It proved that investing according to the mood of the market would have created an annual return of 50 percent. Already burned three times, though, he still shied away from investing. Then a friend showed him a totally different kind of emotion-driven strategy, one that worked great.

The tactic was amazingly simple: Visit the Internet and check out the various stock message boards. Look for hype. Buy a stock that has people excited, and hold it until waves of hype begin to build around another stock. Latch on to that new stock, and repeat the process.

Peterson soon found that he could make that strategy work like gangbusters. In 1999 his profits were 800 percent. In the first months of 2000 they were close to 300 percent.

In May 2000 he left Texas for a psychiatry residency in California, near the epicenter of Silicon Valley. The program was so demanding that for several months he didn't even think about trading. But gradually he returned to wondering about the emotional basis for how people make their financial decisions. Late in 2001 he met a psychologist named Frank Murtha, who was researching gambling and who, like Peterson, was also fascinated by the stock market's psychological mysteries. They combined forces to launch Market Psychology Consulting, a counseling service for people who made big-scale financial decisions, to help them steer clear of emotion-driven mistakes.

When he'd finished his psych residency, Peterson plunged into neuroimaging lab research at Stanford with Brian Knutson. Knutson proved in 2000 that simple pictures of cash caused more intense brain reactions than did gruesome photos of crime and accident victims.

Even more extreme brain activations happened in Knutson's lab when research subjects were actually *offered* sums of money. Dopamine, the brain's reward chemical, the chief feel-good molecule our bodies manufacture, came surging into their synapses as if the subjects were freshly injected drug addicts. For the first time, researchers saw visual tracings of the reasons why money triggers wild extremes of behavior, from the best that people are capable of to the worst. Research with fMRI also validated what Peterson had deduced back in Texas: Money stirs storms of truly primitive emotion, clouding our reasoning ability. We honestly believe in the importance of rational decision making, especially where money is involved. But—even though we seldom consciously know it—our emotions dictate our financial choices again and again.

This finding throws the *Homo economicus* basis of conventional economic theory into a spin. But it had already been proven to be wobbly just two years earlier, in 1998. Some of the best, brightest, and most successful financial minds in the world, by following what they believed was rational decision making, sparked a disaster that nearly triggered a downturn on the magnitude of the crash of 1929.

John Meriwether, a legendary bond trader at Salomon Brothers, founded Long Term Capital Management in 1993 around a team of financial superstars, including two who held Nobel Prizes. By 1998, LTCM was a financial colossus.

The company was based on a very simple idea and backed by the latest supercomputers. Its strategy was to find arbitrage opportunities—flip-it-fast investments that earn profits without bearing risk—and then to immediately borrow huge amounts to exploit the opportunities they had found.

Arbitrage requires discovering something you can buy someplace and then immediately sell at a profit in another. It's an excellent trick, and often portrayed as glamorous. In reality, though, arbitrage is excruciatingly slow, sort of like panning for gold. The opportunities vanish quickly, so you have to grab them in a heartbeat. Plenty of other prospectors are also working the mountain. But after hours crouched by the data stream, looking for glints, you might get lucky. Sometimes there will be a little nugget, maybe even a sizable lump of gold, under all that numerical sand.

Another piece of LTCM's strategy was leverage. With their great academic reputations and mounting successes, the principals convinced some of the world's most powerful banks to cut sweet deals on loan rates. They could fund large positions at superlow cost.

The plan ran astoundingly well, and almost overnight LTCM was one of the world's major financial players. It generated returns of almost 50 percent in its first two years, truly astronomical results. Within four years it had acquired nearly $5 billion in capital, plus assets that it could leverage worth nearly $200 billion. All the world's leading investment banks, including Goldman Sachs, JPMorgan, and Merrill Lynch, wanted some of that gold. They put some miners of their own to work the stream. Increased competition suddenly made those already-elusive arbitrage opportunities even harder to find. Profits slumped. LTCM's investors got edgy.

Two powerful emotions, greed and fear, began to distort thinking inside some of the sharpest financial minds in the world. LTCM's wizards decided to up the ante: Since arbitrage opportunities were now scarce, they gambled bigger sums on the few they could find.

According to their proven models, and to prevailing preneuroscience economic theory, any bond that's priced very low will attract buyers. Following this logic, LTCM voraciously bought up contracts in Russian rubles, which were then available at what seemed to be rock-bottom prices. To hedge against a possible drop in the ruble they also purchased thousands of contracts in Italian lire, Brazilian reals, and Japanese yen. Their strategy was that any investors who pulled money from Russia would then chase after other investment opportunities, which would send the value of LTCM's Italian, Brazilian, and Japanese contracts skyward.

Altogether they backed this bet with $1.4 trillion, most of it borrowed.

Russia's government had been under pressure to devalue its currency for a long time but kept claiming it wouldn't. On an August day in 1998 it finally caved, sending the ruble into a nosedive. LTCM didn't panic: It was hedged in those other currencies. However, what the speculators didn't realize was that their emotions had lured them into a really bad gamble. Their reasoning was clouded by an enormous emotional blind spot; $1.4 trillion worth of fear had tweaked their perceptions so much, they couldn't see the possibility of a global disaster.

Here's what went wrong: When there's a lot of adverse news in the air, bad juju arises in our financial reasoning like vapors from a swamp. Investors can get so skittish that the lower prices they see don't look to them like opportunities. They look like Detour signs. Instead of thinking "What a deal!" they start worrying that something has to be *wrong*. So, contrary to the *Homo economicus* economic theory still taught in universities today, the investors scared off by troubles in Russia didn't turn around and buy up currency in a different country. They cut out risk entirely. Investment evaporated in Russia, and simultaneously did the same in Italy, Brazil, and Japan. Investors ran for emotional shelter, the safest possible investment: low-risk, low-yield U.S. Treasuries.

During five weeks from August into September, LTCM lost a *daily average* of $300 million. In a single day it hemorrhaged a phenomenal $553 million. As it reeled toward bankruptcy, its descent pulled the banks behind its "risk-free" arbitrage opportunity—including Merrill Lynch, Goldman Sachs, Salomon Smith Barney, Chase, UBS-Swiss Bank, Bear Stearns, and JPMorgan—into a harsh riptide. If LTCM hit bankruptcy, they would have been left holding the most expensive vapor trail in world history up to that point. With an enormous financial disaster looming, the U.S. Federal Reserve jumped in to organize a worldwide rescue effort. Fourteen major banks poured in $300 million each, creating a fund of more than $3.5 billion, and LTCM humbly swam back to shore.

The sub-prime mortgage crisis that erupted in the summer of 2008, causing Lehman and many banks to fail, shared many similarities with LTCM.

Both of these stories deliver lessons that reverberate on the same wavelength neuroscience pioneers were beginning to discover about the same time that LTCM went spiraling. We have powerful emotional reactions to events that happen, and also to events that we believe are going to happen. Powerful areas of emotional processing in our brains get fired up, and when they do our whole basis for decision-making shifts, suddenly and radically.

Neuroscience has the potential to eliminate these insane bubble-and-bust cycles in the future. To ignore what neuroscience is learning about our financial brains is like chain-smoking around a leaky gas pump. For example, how did the sub-prime crisis develop, and proceed so far despite many warnings from experts with more rational perspectives. The economist George Loewenstein of Carnegie Mellon University, who also has vital things to say a few pages forward, recently told *Forbes* that brewing the disaster took a combination of stupidity and wrongly organized incentive structure. Ratings agencies were telling people that cutting risky mortgages up into numerous bits would create safe investments. But the same agencies were being paid by the same people who issued those hazardous mortgage-backed securities. The all-too-human result was that the agencies, thought to be reliable guides in the financial jungle, had profit-based reasons to believe in a bad concept. Cutting up the risk-laden mortgages didn't make anyone's investment safer; it just spread the misery around to a greater number of hands.

Daniel Kahneman of Princeton has focused his research for a long time on how we make decisions in uncertain situations, not only big ones like the LTCM debacle but also everyday choices. He and the late Amos Tversky, collaborating in the early 1980s, formulated what they called Prospect Theory. It states that decision making under risk can be viewed as a choice between "prospects" or gambles. We try to figure out the best prospects by basic principles of probability. But, as Kahneman and Tversky discovered, human judgment often takes shortcuts that toss those principles out. So before brain imaging made it easier to look at how humans make judgments, Kahneman was already striving to learn as much as possible about how our emotions impact financial decisions. He and his colleagues helped set the stage for neuroeconomics to arrive.

Daniel Kahneman shared the 2002 Nobel Prize for Economic Science with Vernon Smith of Chapman University. Though they work independently, and to some extent competitively, they have created a large body of neuroscientific evidence that is now accepted as bedrock truth.

Kahneman was among the first to borrow insights from psychological research about how people make judgments and decisions. Smith has been a leader in conceiving and performing lab experiments in order to prove or disprove various economic theories.

Early on, Smith and Kahneman validated what Peterson was learning independently with the up-and-down performance of his intuitive investment tactics: Our personalities contain behavioral biases, which often spin our financial decisions in nonlogical directions. The more pressure there is in a given situation, the more ambiguity and stress, the more these illogical quirks will elbow in and make a grab for the mental steering wheel. So the more distressed we feel, the more our own emotions become saboteurs working deep inside our defensive perimeters.

A classic experiment proves the point. Ask a group of subjects to suppose that they are in charge of a difficult situation. Six hundred people have just come down with a very dangerous disease. The subjects must make a decision, and there are only two choices. The first choice would save the lives of two hundred of the endangered people. The second choice promises a 33 percent chance of saving everyone, but also a 66 percent chance that everyone will die.

Time after time, people favor the first option.

Now try an experiment that is almost identical, except this time the subjects are told about their choices like this: One course of action will result in the deaths of four hundred people. The other offers a 33 percent chance that no one will die, and a 66 percent chance of all six hundred people dying.

With the scenario presented this way, most people take the second choice. But the choices in both experiments are mathematically identical. Why do most people favor choice number one in the first experiment, then switch to choice number two in the second? It's all about emotions.

In the first experiment, both choices are stated as a positive. In the second experiment, they're placed in a negative frame. Because it requires a life-or-death choice, this experiment excites emotions powerful enough to distort thinking. It shows that *Homo economicus*, the individual that classical economics presumes to exist, who always goes for the rational best choice and turns his back on the rest, is a beast as mythical as the unicorn.

How does this knowledge play out in practical everyday financial decisions? As the experiment shows, our emotions impel us toward the sure thing. The thought "I'm going to save two hundred of these poor souls" pulses with a positive emotional charge. "Four hundred people will have to die" does not.

In fact, we love certainty so much that we overvalue it. In the capital markets, people will usually accept lower returns if they believe they are avoiding risk. They tend to invest first and foremost in money funds, or whichever other assets come as close as possible to being a sure thing.

For example, in the thirty-five year span between 1972 and 2007 an investment form known as cash-equivalent investments—the safest of all in standard terms—has averaged a very steady 6.7 percent yield. They haven't had a down year. A slightly different investment—five-year U.S. Treasury bonds—has suffered three money-losing years over the same thirty-five years.

If the thirty-two years of wins are averaged against the three years of losses, five-year Treasury bonds have pulled in annual returns of 8.8 percent. So investors who have taken the cash-equivalent route have been losing 2.1 percentage points, giving up almost a one-third increase in profits per year to avoid a very small chance of losing some in a given year.

Is that rational? No. But it's totally consistent with our emotion-driven thinking. However, there is good news that I'll explain up ahead. It is possible to learn ways to detect those self-limiting, profit-robbing emotions, and then override them so we can consistently make more money from investments.

The flip side of the "sure thing" coin is that whenever something important is at stake, we will try to avoid even the mere *chance* of losing. The pain of losing money operates in one brain region; the pleasure of acquiring it operates in a different one. The pain circuitry hits us with far more intensity.

When the dot-com bubble burst in March 2000, and the NASDAQ dropped from its index peak of 5,048 to less than half of that within one year. Investors remained fearful of the stock markets for a long time. They kept on sluicing capital into bond funds until the middle of 2003. Unfortunately, the interest rates for bond funds were so low that inflation and taxes ate up all possible returns. Meanwhile, even though the stock market was signaling a nice rebound, and there was plenty of encouraging predictive information to be seen, the taste of recent stock-market losses remained bitter. Investors listened to their fears, ignored the facts, and let big opportunities pass by.

Kahneman's work in recording the loud emotional hum inside of our economic brains has led him to team up with three other prominent thinkers in recent years. They are Daniel Gilbert, professor of psychology at Harvard's Social Cognition and Emotion Lab, Tim Wilson of the University of Virginia, and the economist George Loewenstein of Carnegie Mellon. The group calls its field "affective forecasting," which you could translate as "predicting the emotional weather."

The team has been methodically researching two intriguing questions about human nature. First, how do we predict what is going to make us happy? Second, how do we convert these thoughts into decisions?

Along the way the researchers discovered an even more fascinating question: How do we actually *feel* when the results of a financial decision—whether good or bad—truly hit home?

According to their findings, practically everything we do in life is based on mechanisms our minds have built for predicting how the outcomes of our decisions are going to make us feel. These mechanisms are what prod us to buy stuff we think will make us happier, and to attempt career moves that we believe will deliver a bigger taste of our life's potential.

As is typical of homemade devices, our mental mechanisms are afflicted with numerous bugs, quirks, and design flaws. Nevertheless, we have to count on them. They're the only predictive mechanisms we've got.

Stop a moment to consider your own personal predictive machinery. Has it ever led you astray? What if you could have it debugged? What if from now on it consistently produced nothing but right answers? You'd have total immunity from buyer's remorse, you'd know whom to fall in love with, and you'd be fully insured against the pain of career-choice blues.

Kahneman and his cohorts are working on it. They've compiled an impressive number of experiments to pin down the pitfalls we have to jump over in order to make our predictions fault-free. One thing they've established so far: We generally don't understand what it is we *really want*. That's what makes most of us so inefficient when it comes to fixing our lives, and also what keeps self-help books zooming up the bestseller lists. In conventional economic terms, we need help in maximizing our utility.

James Thurber wrote a story about someone who scorched her hand over her kitchen stove on purpose. She wanted to see if the new brand of salve she'd just bought at the county fair was effective. It turned out that it was only so-so. That's kind of how our minds operate. They keep steering us into a wide, discouraging gap between the results we expect and the reality we meet. Boat owners have a rueful saying about this phenomenon: "The two best times you'll have with your new boat are the day you buy it and the day you sell it."

That saying applies to practically all of the toys and baubles we buy in pursuit of happiness. Gilbert calls this gulf between expected enjoyment and experienced reality the "impact bias."

As Loewenstein explains, "Happiness is a signal that our brains use to motivate us to do certain things. And in the same way that our eye adapts to different levels of illumination, we're designed to kind of go back to our happiness set point. Our brains are not designed for trying to be *happy*. They are designed to regulate us."

Loewenstein uses the term "empathy gap" to describe the difference between behavior when our emotions are highly charged, which he calls "hot" states, and when we're calm and rational, which he calls "cold." Our minds operate like the old Jerry Reed hit from the seventies: "When you're *hot*, you're *hot*, and when you're *not*, you're *not*." We all have tendencies to mania, to some degree or another. We all find ourselves in states of mind where we're prone to make ill-advised choices because we strongly believe, just at that time, that they're going to turn out fabulous.

"These kinds of states can change us so profoundly," says Loewenstein, "that we almost become another person."

Wilson adds, "We don't realize how quickly we will adapt to a pleasurable event and make it the backdrop of our lives. When any event occurs to us, we make it ordinary. And through its becoming ordinary, we lose our pleasure."[1] Mark Twain had a similar thought around a century ago: "As soon as the novelty is over and the force of contrast is dulled," he wrote, "it ain't happiness any longer, and you have to get something fresh."

I expect the future will bring us some neuroscience-based interventions that will free us, at least to some degree, from the discouraging effects of impact bias and the empathy gap. We simply need to correct our tendency to overestimate. It may be that we get foolish in our pursuit of happiness because we need more of it in our lives. There are many possible pathways, including another new hybrid discipline you'll read about in chapter 7, called neurotheology. In terms of neuroeconomics, we may find help in neurosoftware programs to retrain our brains, neurofeedback solutions to provide immediate information on our precise mental state, or a drug tuned to make itself active only during peak events. Being aware of your empathy gaps and impact biases, and having ways to override them, would just be the starting point. You'd have to learn to recognize what it feels like at the times when your thinking, especially your ability to gauge value, is in danger of getting turned sideways. A piece of electronic equipment, something you could tap into as easily as a heart-rate monitor, could warn you. Monitoring our emotional states will reduce our overestimation tendencies, so our expectations and resulting realities will fit with a tighter seam.

People are already working on this possibility.

Interventions like these could give us a greater sense of being in control of life and its directions. We could know whether pulling up stakes and creating a new life in a different part of the country, or changing our career, or even just driving a different car, would repay the financial and emotional expense required. And suppose that—after neuroeconomics research has validated the wisdom and benefit of altering some laws—when you make a big purchase you'll be given a certain amount of time to assess whether you've gone overboard. Buyer's remorse could become a thing of the past, and so could the seller's version of the same kind of regret.

"A life without forecasting errors would most likely be a better, happier life," says Loewenstein. "If you had a deep understanding of the impact bias, and you acted on that understanding, you would tend to invest your resources in the things that would really make you happy."

Financiers are, of course, among the first people who will jump on these developments as soon as they become available. Some are already playing with the potential of owning stronger and more lasting advantages in trading performance through neurotechnologies like brain imaging and neurofeedback. Exploring how to apply these medical technologies to economics and finance is how the leading-edge scientists came up with the discipline of neurofinance.

Neurofinance is the newest shoot on the family tree that includes behavioral finance, behavioral economics, and neuroeconomics. Behavioral finance and behavioral economics use concepts from the behavioral sciences, especially social psychology, but aren't as closely linked to neurotechnology.

Neurofinance researchers chase three different holy grails related to improved trading performance. The first is to identify which physiological traits affect our trading behavior. The second is to see how these traits lead to either success or failure. The third is development of powerful new tools, technology, and training methods to make sure that profits are maximized.

Neurofinancial research has established that our evolutionarily conditioned psychophysiology prevents us from rational behavior, and that we all have varying psychophysiological makeups that affect our individual behaviors in several different ways. These individualities, in turn, greatly affect both our ability to make rational decisions and our success as financial market operators. This finding drives a stake through the heart of the long-prevailing Efficient Markets Hypothesis, known as EMH for short.

EMH, along with its antecedent, Utility Theory, continues to drive most modern financial decision making. Both of those older schools of thought believe that people usually act out of rational self-interest and make their economic decisions accordingly. EMH and Utility Theory are so widely held, and still so persistently taught, that most financial professionals consider them to be gospel. But these theories not only go against the grain of current research; they insult the intelligence of anyone who has ever thought much about how things happen in real life. Realizing that we're all ditsy sometimes about money decisions is literally a no-brainer.

The question is, once we accept the fact that we aren't entirely rational about money, what can we possibly do about it?

If practitioners of neurofinance have their way, traders in the near future will work with the aid of unobtrusive monitors that continuously scan their brains and test their blood chemistry. This idea may seem strange, perhaps unethical, and possibly stolen from a lost fragment of Aldous Huxley's *Brave New World*. However, not so long ago it seemed just as far out to test the lactate threshold and VO_2 max (maximal oxygen uptake, or aerobic capacity) of athletes. Those tests are now routine for swimmers, runners, and other endurance athletes because they offer insight into each performer's range of natural ability based on heart size, aerobic capacity, and other biological characteristics, and they also provide crucial feedback before, during, and after performance.

A trader is a kind of financial athlete with a package of inborn skills and specialized training, always seeking ways to jack up his or her personal best and break out from the pack.

As long as we're toying with sports metaphors: Imagine you own a professional sports franchise, or that you're coaching a team that must face the world's best competitors at the next Olympic Games. You need to wring supreme performances from your team. You're likely to pursue every legal means at hand. (But not the illegal. We're assuming that you're also ethical.) Sooner or later, you might pay a call at Sandia National Laboratories, just outside Albuquerque, New Mexico.

Since 1949 Sandia has had the job of developing technological solutions to support U.S. national security and to counter national and global threats. That original mission is still Sandia's first priority. It provides engineering design for all nonnuclear components in America's nuclear weaponry. But Sandia also performs an amazing variety of other national defense R&D, energy, and environmental projects. Its Advanced Concept Group is working right now on the extreme technological cutting edge of how to improve human cognitive performance.

No one yet understands exactly why groups of people will perform optimally at certain times and below par at others. But even without a coherent theory to guide them, Sandia researchers are trying to identify the characteristics and conditions that set the stage for optimum team performance.

By cobbling together some different pieces of commercially available equipment, they have created a device they call an anthroscope—literally a "human watcher." The components include accelerometers to measure motion, face-recognition software, EMGs to measure muscle activity, EKGs to measure heartbeat, blood volume pulse oximetry to measure oxygen saturation, and a Pneumotrace respiration monitor to measure breathing depth and rapidity.

They've named the anthroscope PAL, which rhymes with HAL, but hopefully their invention is more benevolent than the astronaut-killing computer of that name from 2001: *A Space Odyssey.*

PAL can simultaneously monitor a subject's perspiration and heartbeats, read his facial expressions and head motions, and analyze the rise and fall of his voice. It then correlates all the numbers and delivers a running account of how together that person seems to be at the moment. PAL can transmit this information to others in a group, so everyone can know how well everyone else is likely to perform at a given point in time. Who, to ride the sports metaphor a little longer, should be the one to take the clutch shot when the game is on the line.

Sandia's original idea was to use PAL in nuclear subs and silos and in air traffic control towers, so that everyone on a team could know which individual among them was most capable of making optimal decisions at a crunch time.

With some slight adapting, Sandia could turn its anthroscope into a neurofeedback apparatus that feeds you (the team owner or coach) real-time neurobiological information from your athletes during learning situations. That data would enable you to improve their performances and outcomes by altering their behavior in midstream, giving you a monster-sized competitive advantage.

It's already been proven. In an early PAL test, five people interacted as a team in twelve sessions. The first thing they noticed was that information coming in via biofeedback helped them remain in lower arousal states, less keyed up in general. That factor alone improved their teamwork, and also led to better leadership in longer collaborations.

A lowered arousal state makes possible what Buddhist meditators call "the effortless effort." When you're in a lowered arousal state, you stay aware and alert, ready to be effective but not burning up a lot of mental or physical energy. It's like the state that athletes call "the zone." When a basketball player is in the zone, his shots swish through the basket as if guided by radar. A runner who's in the zone feels as though the high hurdles are politely tucking themselves out of the way as he or she strides over. Touchdown passes are caught, tennis volleys returned, low inside pitches smacked over the fence without any strain. The mind just floats, and athletic training takes over.

The opposite state, overarousal, is that sweaty-palmed, brain-churning condition called "fight or flight." When we are overaroused, adrenaline flows. There's an energy surge. Most times, it isn't appropriate in modern life to run away screaming or to club someone who is standing in front of us. But our minds are thinking "Do something *now*!" A little adrenaline can be helpful, which is why people love the boost they get from their morning coffee. Too much adrenaline is almost a guarantee of costly misinterpretations and outright mistakes.

The next logical step for neurofinance, of course, is using these technologies and knowledge to boost the performance of traders. This is exactly what Andrew Lo has been doing for the past few years at the Massachusetts Institute of Technology.

Lo was born in Hong Kong but has been a New Yorker since age five. He teaches finance and investments at MIT's Sloan School of Management, and also serves as director of the MIT Laboratory for Financial Engineering, a research center that uses computational models to study financial markets.

In "Psychophysiology of Real-Time Financial Risk Processing," a paper published in the *Journal of Cognitive Neuroscience*, Lo and his co-investigator, Dmitry Repin, looked at emotion in the high-stakes, high-pressure world of professional securities traders. These traders are in the elite category of financial athletes. Individual investors are often guided by irrational feelings such as overconfidence, overreaction, herd mentality, loss aversion, fear, greed, or simply excesses of optimism and pessimism. By contrast, professional traders seem to face the pressures of minute-to-minute transaction opportunities with the easy self-assurance of Cool Hand Luke. By both training and inclination, they are believed to be about the most steady decision makers in the economy. It's commonly believed that they work purely on the basis of intellect and rational analysis.

When Lo and Repin set out to test this notion, they found that even the most seasoned and tested professional traders can get pushed around by their own emotions. Using their own version of an anthroscope, they measured and analyzed autonomic nervous system data from ten professional traders working in foreign exchange and interest-rate derivatives for a major Boston financial institution.

The traders were divided into two groups, based on their business experience levels: five traders with high experience, five with low or moderate experience. With the anthroscope watching, they made decisions in real time during live trading sessions. Attached to their face, hands, and arms, the subjects wore electronic sensors that continuously monitored and recorded their skin conductance, heart rate, respiration, facial and forearm muscular activity, and body temperature. The sensors were wired into a small control unit worn unobtrusively on each subject's belt.

A fiber-optic connection led from the control unit to a laptop computer, where biofeedback software analyzed the real-time physiological data so it could be compared to the ebb and flow of the real-time financial data on which the traders were making their decisions, allowing researchers to compare the market events and the simultaneous human reactions side by side.

Three kinds of events were tracked. The researchers chose to focus on deviations, trend reversals, and periods of increased market volatility—exactly the sort of high-drama occurrences that you'd expect to generate heightened arousal, even among the most experienced traders. Data was collected before, during, and after each event.

The market data consisted of prices and bid/offer spreads in thirteen foreign currencies and two futures contracts. Lo and Repin were interested in seeing how different the more experienced and less experienced traders would be in their reactions, and also to test whether responses by the traders would change depending on the kind of financial instruments traded.

Not surprisingly, Lo and Repin saw each trader's autonomous nervous system register excitement each time there was a dramatic market event. Rookies reacted with slightly more intensity, but even the most highly experienced traders showed very significant physical responses. Emotions were proven to be a vital and inescapable factor, even among the most rational investors. The big difference seemed to be in how those emotions get handled.

Interestingly, the most successful traders often seem to lack the ability, or even the need, to articulate how they reach their decisions. Lo and Repin think that these traders simply have the right kind of innate emotional mechanisms for forming intuitive judgments, and that when they're asked to explain those decisions, they construct a rational framework after the fact.

This conclusion, that the traders get emotional but just somehow know how to ride those feelings confidently anyway, fits well with current theories in cognitive science. Cognitive scientists believe that emotion is the basis for a reward-and-punishment system, one that spurs us to naturally select evolutionarily advantageous behavior. From an evolutionary perspective, emotion is not a hindrance but rather a powerful adaptation. Its purpose is to improve dramatically the efficiency with which humans and other animals learn from both their environment and their past.

Emotion, then, is actually a major part of the financial professional's tool kit. It is something that significantly determines the evolutionary fitness of a financial trader. Unsuccessful traders, after a certain level of losses, are generally eliminated from the population. So having emotions, if you also know how to keep them in perspective, is part of a winner's total package. It's a conclusion that makes sense of the steps that Peterson had to make as he worked his way up toward investment success to the point that he is now putting neuroeconomic research to work in managing a $50 million hedge fund, MarketPsy Capital. In 2008 the fund had gains of nearly 40 percent while the Dow Jones Industrial Average dropped over 35 percent.

Neurofinance researchers believe their field has virtually limitless potential. "The brain scientists are the wave of the future in the financial world," Kahneman says. "Whether you're in academia or in the investment community, you'd better pay very serious attention to them."[2]

David Darst, who is chief investment strategist for the $700 billion individual investor group at New York–based Morgan Stanley, implicitly agrees. "One day," he says, "brain science may help money managers spot shifts in investor sentiment."

Brian Knutson at Stanford is one of many neuroscientists who expect that traders will use highly targeted psychoactive drugs to help them be more profitable, more consistently. Today's psychostimulants, which are very raw and side effects–laden compared to the drugs that will be developed with the aid of brain imaging, are already becoming popular. Chapter 8 will explore some surprising facts around the vast numbers of students and young professionals already taking psychostimulants to improve their academic and job performance.

Knutson's prediction also jibes with increased public awareness about brain chemistry, knowledge that has multiplied in the years since Prozac came on the scene. Prozac and similar SSRI (selective serotonin reuptake inhibitor) drugs have not just revolutionized the treatment of depression; they've also profoundly changed how we view the mind. Most people now recognize that chemistry drives their brains, moods, and behavior, and that improving their brain chemistry will help them function better.

There will probably be, in the near future, an entirely new wave of financial trading systems that leverage advantages derived from neurotechnology. First, imagine you have breathing techniques or drugs to reduce the anxiety behind trading jitters while keeping you mentally sharp. (Sophisticated breathing techniques have been around for millennia: Now we can test what types work best, in what specific kinds of situations, via fMRI.) Now combine those tools with real-time brain scanning, and with neurofeedback software that quantifies how your brain is acting at this moment and compares that information with the way your brain acted during previous trades, both the successful and the unsuccessful ones.

The whole package would enable you to know if you were in the zone, ready to make smart choices without strain. You'd also know if you were way off, and ought to rest on the bench, relaxing until better emotional weather fills your personal skies.

The precursor to this neurofeedback system is being developed today, at a Menlo Park, California, company called Omneuron. Omneuron is working with Stanford University to complete development of a new technology called real-time fMRI, which will be used to train patients in pain management techniques. A very recent invention, real-time functional magnetic resonance imaging makes it possible for an individual to look at which specific brain areas are activated when he or she thinks about a specific event or focuses on a particular task. In pain management tests, patients have shown that a thirteen-minute session with Omneuron's device helps them learn to control activity in different parts of their brain and alter their sensitivity to painful stimuli.

The patients are able to watch a screen where they see in real time the activity of a part of their brain involved in pain processing. As they watch, they practice mental exercises to decrease this brain activity. The screen tells them how well they are doing. In a paper published in the *Proceedings of the National Academy of Sciences*, it was reported that eight patients with chronic pain that couldn't be adequately controlled by conventional means experienced a 44 percent to 64 percent decrease in pain after the training, a benefit three times larger than that reported by a control group. Patients who showed higher ability to control their brain activity showed the greatest benefit in pain reduction. In addition to managing pain, this technology is also being used for addiction and many other brain and nervous system ailments. The same system could be used today by anyone who wanted to gain competitive advantage in financial trading. When Omneuron, or some other company, manages to build a smaller machine at lower cost, traders will line up to be on that cutting edge, and they'll generate results that win them bigger and bigger investment portfolios to play with. As the costs of neuroimaging drop and the technology arrives in smaller packages, it will eventually become as invisible in our environment as the copper pipes that deliver the water to our faucets, or the fiber optics that send exabytes of data instantly across the globe. Neuroimaging technologies will become as universal among tomorrow's financial traders as trading screens and mobile phones are today. The first generations of the technology may be a bit bulky and even somewhat obtrusive. But with billions of dollars on the line each second, intelligent traders around the world will grab for the first-generation devices in order to get as far ahead as possible, knowing that lower-cost second-generation devices will spread the technology further, putting it into the hands of more of their competitors.

Neurotechnology represents the next form of competitive advantage, beyond information technology. Enabling a higher level of productivity, it creates what can be called the neurocompetitive advantage. Just as workers today leverage information technologies for competitive purposes, workers in the Neuro Revolution will pursue neurocompetitive advantages.

As something incredibly valuable on their own, and as a possible complement to sophisticated upcoming neurosensing technologies, neuroscience is now bringing new highly targeted pharmaceuticals onto the horizon. I call these neuroceuticals. As more people live longer lives, and as global competition intensifies, many of us will turn to regulated neuroceuticals as an advanced set of tools to help us survive and succeed. Cogniceuticals will be a class of medicines for increasing memory retention. We will use emoticeuticals to decrease stress, and sensoceuticals to add a meaningful pleasure gradient to our various activities.

At any given moment, our neurochemistry is what chiefly determines how we feel, and how we perform in the world. Neuroceuticals will allow those who choose to leverage them the ability to compete at peak levels, and in some cases enjoy life itself more, by helping them have more optimal neurochemistry more of the time. But expect neuroceuticals to create plenty of social controversy. Should they only be allowed for people with illnesses? Or should we let normal healthy people have them too? These questions bring up many ethical dilemmas, which we explore in later chapters. For now, just imagine what the world would be like if just a small group of financial traders—in Dubai, Hong Kong, Mumbai, or your hometown—suddenly had access to neurosensing technologies and neuroceuticals.

This sort of enhancement will create a new, more efficient, and hypercompetitive playing field. That field will be dominated by those traders who possess the advantage of predictive capacity based upon their continuously shifting neurobiology, just as surely as banking and other financial professions became dominated by the companies that were quick to take advantage of canals, railroads, telegraphs, computing power, and so on. The growing number of people who are pushing neuroeconomic and neurofinance research ahead will soon bring more ideas to light that can raise the profitability of multibillion-dollar transactions and steadily create wealth. We will soon see masses of financial professionals rushing to take advantage of each subsequent breakthrough. Nothing will inspire a person to try out new ideas and technologies more than believing that big stacks of money are waiting just ahead, right at that beautiful spot on the horizon where the present meets the future.

FIVE

TRUST

It is an equal failing to trust everybody and to trust nobody. —Thomas Fuller

What you need is Love Potion Number Nine.
—Jerry Leiber and Mike Stoller

Money in hand is wonderful. But there is a different kind of wealth, even better than money. It's called "social capital." When you've got it, you're surrounded by plenty of the conditions that create wealth, and you can trust that those same conditions will generate a lot more riches in the days to come.

Social capital includes every single thing that helps a society run smoothly. It is the sum of everything tangible and intangible, institutional and cultural, that helps keep our social and economic collective life in a zone. Social capital accumulates around any factor that boosts the common good and minimizes social and economic friction—anything, in other words, that makes teamwork feel more consistently rewarding than selfishness.

If a country has high social capital, prosperity is either well established already or about to arrive soon. Where it is low, poverty stays powerfully entrenched. Neuroeconomists believe the more we understand social capital, the more success we will have creating a strong, stable global economy, where the available prosperity reaches far more members of humankind. And here is where a rare, subtly acting, yet profoundly powerful neurohormone called oxytocin enters the picture.

Oxytocin is a naturally occurring hormone. Because it is built of nine amino acids, scientists sometimes jokingly call oxytocin "Love Potion Number Nine Amino Acids" or "the cuddle hormone" or, a little more seriously, "the affiliative hormone."

Oxytocin gets made in the hypothalamus, then released into our bloodstreams by the pituitary gland. It helps us like each other, and makes us value being in social situations where trust, kindness, and love are mutually exchanged. You can think of it as the neurochemical basis of love, of all human attachment and bonding, and therefore of cooperation and teamwork on all levels.

You can also think of oxytocin as providing a tremendous competitive advantage, and as something that potentially will make life in the coming neurosociety more pleasant and humane than today's existence.

We can synthesize oxytocin, but unfortunately it happens to evaporate quickly in our systems. If we wanted it full time, we'd have to go around rigged up with devices like those ridiculous helmets some sports fans wear—two beer-can holders on the sides, tubes feeding into their faces. It makes much more sense to go the natural route. That means learning how to encourage our own natural oxytocin, and figuring out what kinds of conditions will inspire everyone's hypothalamus and pituitary toward a more consistent flow of the magically effective stuff.

Current research, made possible by brain imaging, is striving to learn how we will make this happen. Oxytocin has been around for millions of years, at least since the dawn of civilization—a development that oxytocin made possible—benevolently affecting the behavior of humans and many other mammals. It wasn't actually discovered, isolated, and synthesized until 1953. Vincent Du Vigneaud earned the Nobel Prize in Chemistry for his efforts.

When women give birth and when they breast-feed, they experience a strong release of oxytocin. So do people of both sexes when they experience orgasm.

But releases also occur in other meaningful situations. They may involve smaller amounts of the hormone, but they hold enormous sway over our behavior. They induce feelings that are so positive, we will try again and again to duplicate the circumstances in which those feelings came to us.

Whenever you experience a signal that tells you the person you're interacting with is trustworthy, you get a small surge of oxytocin. Any positive social signal, whether small or magnificent, puts a trickle of the hormone in your system, to make sure you both enjoy and remember the experience. Even a microscopic droplet is likely to induce more trust. So the presence of oxytocin in your bloodstream can be considered a physical signpost of empathy, a proof, in other words, that you are successful at understanding and relating to others.

I'm all for empathy, not to mention all those other good feelings oxytocin induces. I've taken the invitation of Paul Zak, director of the Center for Neuroeconomic Studies at Claremont Graduate University. He is the leading researcher of oxytocin in humans, and I'm going to experience oxytocin in his lab. On a sweetly sunny December day I drive a few miles east of Los Angeles. There, on the edge of the tree-lined campus, in a pint-sized 1920s bungalow converted to research offices, I will receive forty puffs of oxytocin mist in my nostrils. About 10 percent will enter my brain. The remaining 90 percent will find receptors throughout my body.

Hundreds of subjects have already done the same in experiments conducted by Zak and his colleagues.

Like many neuroeconomists, Zak believes that we need to increase our understanding of the brain mechanisms associated with prosocial behavior. He sees potential that this kind of knowledge will lead us forward as a society, help us evolve a world that naturally calls forth our best characteristics and helps us find more contentment. He imagines that neuroscientific principles will be used to design better public policies; to create better human environments in homes, schools, and workplaces; and also to help us find more ways to naturally increase beneficial, prosocial behavior while reducing destructive antisocial tendencies.

Zak's scope of concern is wide enough to include international trade. In *Moral Markets*, a book he edited, several of the contributing writers assert a belief that market exchange can make us more virtuous. We think of competition as a driving force that advances the greedy and the selfish, but self-interest can place competition inside a framework of cooperation. It's true that competition is a basic part of human nature. We spend billions every year supporting young millionaires, and the billionaires who pay their salaries. Our heroes in baseball, hockey, football, basketball, and other sports put on uniforms and savage identities to improvise dramas, inside (mostly) sets of rules, about pushing another tribe off the map or raiding its territory. But even though we worship our competitive urge and pin medals on the top dogs, cooperation ultimately swings the most weight with humanity. We all belong to teams in life, from families to friendship groups to businesses to nations. We have to blend our efforts. We have to maintain key relationships, and those connections need peace to survive.

When trust is scarce between nations, we start planning for war. Sometimes, to avoid the costs of war, we first use the international strongarm tactic of cutting off trade to gain diplomatic leverage. It's frequently just a prelude to war. Michael Shermer, a Claremont economics professor and contributor to *Scientific American*, cites the French economist Frederic Bastiat. Reflecting on the strife of the Napoleonic Age, Bastiat wrote, "Where goods do not cross frontiers, armies will."

Shermer's book, *The Mind of the Market*, points out that United States economic sanctions in response to the Japanese invasion of China helped fuel the decision to bomb Pearl Harbor. In more recent times, economic sanctions have exacerbated problems with Iraq, Iran, North Korea, and Cuba. There may be legitimate political reasons for applying economic sanctions, but unfortunately they seem to accelerate the breakdown of trust. With the loss of trust comes the loss of peace.

Zak's work proves the powerful connections between trust, trade, and economic well-being.

Conventional, preneuroeconomic theory predicts that a rational, self-interested person should never trust another person. And yet trust is vital to civilization. Nothing financial can possibly happen without it. To have an economy, to have any kind of exchange, we need to reasonably believe that there are times when we can grant trust to someone and come out okay.

Imagine what the world would be like if there were no such thing as trust, or the cooperation it allows. We'd all have to grow or hunt all our own food, build and guard our own private shelters, stitch our own clothes from cloth spun from the fibers of plants we would have to cultivate, or from the hides of animals we would have to trap, and so on, into total impossibility.

Even if such extreme self-sufficiency could happen, humanity would vanish in a single generation: Babies with completely selfish parents would die immediately.

Zak is known as the "King of Trust." Based on our interactions at several conferences, I can see why. His ability to communicate and connect with others is just as striking as his research into trust is groundbreaking. His ultimate goal is finding ways to make trust thrive in the desert, so to speak, and helping people of poorer nations reach higher ground. He has an infectious enthusiasm about how it's eventually going to happen. In fact, he's so sure about the eventual emergence of the neurosociety, it sometimes seems like he's living there right now.

Zak wheels his family SUV into the narrow driveway in front of his bungalow offices. His license plates announce his enthusiasm up front, in capital letters: OXYTOCIN. He steps out, a tall guy in shiny black cowboy boots, still on the sunny side of middle age and possessing a wall-to-wall smile. Zak leads me into his office, seats me in a chair. Out comes a white plastic bottle with a plunger top. It looks exactly like the ones found on the colds and flu shelf of a drugstore.

But before he instructs me to lean forward so he can have a straight shot at my nasal passages, we cover the preliminaries: Yes, I was just screened by an M.D., and I've signed a permit form. Yes, I realize that about 25 percent of men who receive oxytocin this way experience an erection.

I am a researcher; I am not afraid.

Nevertheless, it's reassuring to learn I'll also probably experience lowered blood pressure and I won't feel "out of it," foggy in the head about my surroundings or about what I'm doing. I'm likely to feel a bit happier, a bit more relaxed, but not so much that I'll want to kiss everyone in sight. The only other risk, besides one chance in four of experiencing out-of-context amorous readiness, is some mild throat irritation and/or red eyes.

I nod that I'm ready, and the plunger bottle heads my way.

The protocol is five puffs in one nostril, pause for a big breath, then five in the other nostril and another big breath, until I've received all forty puffs. Zak proceeds slowly, so as much oxytocin as possible will contact my mucous membranes.

It's a natural hormone arriving at a concentration level that's close to what might be experienced in real life, so the experience starts off mildly enough. At a gradual but unmistakable pace, it becomes more and more pleasant. As for the possible priapic effect, I'm withholding that datum.

Meanwhile, I'm fully aware of my surroundings. I could move around if I wanted to. But it feels fine just to sit . . . right . . . here . . . and to stretch and yawn once in a while. Within two hours about 80 percent of the oxytocin will be out of my system; after four hours all of it will be gone.

Which leaves plenty of time to learn about how Zak got interested in this hormone, what he has achieved so far, and what he hopes to do next.

Over the past four years, Zak's team of economists and neuroscientists has been studying the influence of trust on economic development, and thinking about how we might all collectively keep our oxytocin more free-flowing.

"I got interested in trust," he says, "because I wondered why some countries are poor, and seem like they'll always remain poor. Why can't we fix 'em?"

It all clicked in his mind during a graduate seminar on social capital. "If I'm going to understand social capital," he reasoned, "I should understand trust. It's a good measure of social capital."

Any time you're buying a drill press, a B-flat cornet, or an SUV, you won't spend your dollars unless you can reasonably trust you're buying a good product. As part of the process, you evaluate the person on the other side of the counter. Does he or she seem trustworthy? Or you examine the brochures and the packaging, trying to see if they offer facts and not just hype. If both those factors look good, your decision is easier to make.

People have bought astronomical numbers of new and used items on eBay, dealing with complete strangers, without benefit of face-to-face contact, brochures, or packaging. The enormous success of the ruling online auction company rests on its carefully tended feedback system. You have to trust the honesty of other strangers, that they've fairly reported what their experiences were when they did business with this particular stranger who wants to do business with you. Somehow, you come to believe that the total stranger on the other end of the line will probably pay—or deliver—as he or she has promised. To alleviate your concerns, you click on feedback to get an electronic equivalent of personal contact.

Vast numbers of eBay employees work behind the scenes to keep that feedback system viable enough to promote and sustain high trust levels. They facilitate the resolution of complaints and concerns, and they constantly look for ways to increase the feelings of connectedness among eBay participants.

This is wise policy. Trust is our socioeconomic lubricant. It directly affects everything from personal relationships to global economic development. Economists love trust: It reduces the transaction cost of trading. If trust is present, and reliable, you don't have to be suspicious of the person you're dealing with. This presents a great savings of money and resources as well. You won't feel obliged to spend time and money checking out that person's history and claims. You won't get stuck with poor merchandise, unpaid bills, or fees paid to lawyers. You will wind up saying, "It was a pleasure to do business with you," and you'll really mean it. You and that person have just co-created a win/win situation.

"To understand the relationship between trust and economic development," Zak says, "I began building mathematical models of trust. These did a good job of explaining how increasing the levels of generalized trust in a country correlates to reaching higher standards of living. Poor countries, by and large, are low-trust countries.

"Trust is kind of a summary variable. It picks up all the good or bad things happening in a society. High-trust societies have good social institutions, good formal institutions. They tend to have fairly even distributions of income, higher education levels, higher levels of initial income.

"In 2001 I wrote a paper about trust and economic growth. Lots of people picked up on it and cited it. The World Bank then flew me out to an important meeting. They asked me, 'How do we build trust?' At the seminars, people were asking, 'How do two people decide to trust each other?'

"I'd just say, 'I don't really know. I'm an economist.' "

Eventually, Zak says, he started to feel like a fraud. He had to take on a big job—discovering on a scientific level how people decide to trust each other. He applied for work in Vernon Smith's lab in 2001. Smith is known as the father of experimental economics, and Zak connected with him just one year before Smith and Daniel Kahneman jointly won the Nobel Prize in Economics.

Smith is the researcher who set the standards for reliable laboratory experiments in economics. He has also spearheaded "wind-tunnel tests," where new, alternative market designs—deregulating electricity markets, for example—are tried out in the lab before they get introduced in real life.

In the early 1990s, Smith and Kevin McCabe, his right-hand man, had invented an experiment Zak wanted to know more about. It's called the Trust Game, and it assesses how people understand the intentions of others. Though very simple, the Trust Game is also extraordinarily clever and can be spun into a number of variations. Zak has primarily focused on one called the Investment Game, an exercise that peers into the individual dynamics of trusting and being trustworthy.

Here's how the Investment Game is played: In a room full of test subjects, all participants are grouped into pairs. Each pair of subjects is linked by computers, and subjects can make eye contact with everyone in the room, but none of the subjects know whom they're actually paired with. All of the subjects get $10 just to show up. Within each pair of subjects, one is told that he will be Player One, and can send any amount of his $10—whether all of it, none of it, or some portion in between—to Player Two.

The rules governing the game say the experimenter will triple every dollar sent by Player One to Player Two. Then Player Two gets to decide how much of this amount she will keep, and how much—if any—she will send back to Player One. As soon as both subjects have acted on their decisions, the game is over. A small blood sample is drawn to test oxytocin levels, and the subjects leave.

A "rational" person, the self-interested man or woman of conventional economic theory, should keep all of the money he or she is given. But this rarely happens.

On average, Player One will send $5.

Roughly one-third of the time, Player Two sends more than $5 back. These results contradict standard, preneuroscience economic theory, but Zak's experiments have evoked the same behavior over and over again.

Standard economics relies on the so-called Nash equilibrium, named after the Nobel Prize–winning mathematician John Nash. It predicts zero trust, zero trustworthiness. So now we have a mystery. Why do most people act contrary to the rational-choice model, even with strangers? To make the mystery deeper, subjects can seldom explain the "why" of their decisions. The closest they're likely to come is, "It just seemed like the thing to do."

"The Investment Game is really a shot across the bow of rational-choice economics," Zak says. "These subjects aren't doing this cognitively. They're going by some kind of instinct. Also, whenever people receive a signal of trust, the return data are very, very tight. It's really clear for almost everybody—about 98 percent of the population—that they want to repay the trust."

Zak's training includes economics, general biology, and neuroscience. Biology studies taught him that oxytocin seems to facilitate lots of prosocial behaviors in mammals, especially in monogamous sets in which the males are involved in parenting. The prairie vole is an animal that illustrates well how oxytocin influences social behavior, and it's a frequent choice for studies.

Reasoning that humans are mostly monogamous, or at least serially monogamous, Zak figured that oxytocin probably influences humankind the same way it does prairie voles. He combed scientific literature for oxytocin experiments in humans, only to find none existed. For technical reasons, of which Zak was then blissfully unaware, the effects of oxytocin on people hadn't yet been studied.

"I contacted all the oxytocin researchers in the world," he says. "They all told me, 'We figure that whatever we learn about rodents will apply to people.'

"But even across different mammal species," Zak says, "the distribution of oxytocin receptors is totally different. There's no way I can go from rodent data to understanding humans. So without human studies I can't build the next trust model; I don't have any data to support it.

"So I was stuck. I was frustrated. And yet this stuff is really important. If I can figure this out, I might be able to help alleviate poverty. I may be able to make people's lives better. That's a big deal, right?"

Zak and his colleagues opted to run their own oxytocin experiments on humans. The trick was figuring out how to do it. Oxytocin is in blood and the brain, and the blood and brain releases are coordinated, so they didn't have to do spinal taps on their subjects to gather data. That was the good news. But oxytocin has a half-life of three minutes, and blood degrades rapidly at room temperature. And they needed to find a lab that could analyze the data.

Although Zak didn't know it, researchers at Emory University had just created a highly improved method for detecting oxytocin. It was between one hundred and one thousand times more sensitive than older methods. This new assay had arrived about one year before the Claremont team began work, and very few labs had it yet.

When Zak went looking for a lab to process their experimental blood samples, they picked Emory for two simple reasons: low cost, and its extensive experience with oxytocin research in rodents.

As luck would have it, the improved sensitivity of Emory's test was crucial to experiments with humans. Oxytocin levels in human blood are close to zero most of the time. There has to be some stimulus, or it's not produced. With the old, less-sensitive testing methods, the data might not have revealed anything definitive. UCLA researcher Sue Carter, a pioneer in oxytocin studies with rodents, later told Zak, "You are so incredibly lucky! If you hadn't sent your samples to Emory, you wouldn't have gotten any results."

Zak's team decided to run the Investment Game in an environment where people could not see each other, look in each other's eyes, shake each other's hands, or generally experience contact that would personalize the exchange. That would leave only the intention to trust, without all the other baggage.

When the first-ever running of the Investment Game was concluded and researchers measured the blood of the players, the posttest blood samples showed exactly what the team had hoped to see. Those results have remained consistent ever since. When someone receives an anonymous monetary transfer, denoting trust, oxytocin rises; simply giving people money won't do the trick. The stronger the signal of trust, the greater the increase. And the more oxytocin increases in Player Two, the more likely it is that she will reciprocate generously to Player One.

Participants don't spend a lot of time pondering; their brains simply guide them to be trustworthy. This proves human beings have exquisite innate mechanisms for interpreting and responding to social signals. We are the only mammal that regularly cooperates with unrelated others that we don't even know. And we do it all the time. We're willing to burn time, calories, and other resources to do it. No other species acts like this. When bees and ants cooperate, for example, they do it only in groups that are tightly related genetically. So what the Investment Game results suggest is that our brains are wired to give us quick on/off decisions, without a lengthy reasoning process: "This person is not safe. That one is."

Through that neural mechanism, humans are able to live together, and to have economic lives with wide, even global interconnections.

"This ancient hormone," says Zak, "allowed us to expand from strictly kin-based groups into villages, and then into small cities. That allowed specialization of labor, leading to generation of surplus resources. Suddenly, survival was no longer everyone's full-time job. Learning and other advanced behaviors could flourish."

Another interesting thing about oxytocin: Stress actually increases its release, so a reasonable amount of stress can make a group of individuals bond as a family, a team, a company. This explains the lifelong friendships that can form among people who serve together in the military, for example. But too much stress will make people pull back from the group. Their brains will simply tell them to.

"That's why lunch is the most important meal of the day, from a productivity point of view," Zak says. "That's where you bond with your colleagues. When I used to work at home, after I'd been through a few days of isolation I just had to go seek someone out. One of my neighbors worked at home too. I'd go talk to him. 'Hey, Steve, something good happened to me today.' "

This little anecdote has giant implications. Environmental design should be coordinated with the ways our brains want to work. It should encourage oxytocin release.

Humans do not want to be isolated. It is physiologically and psychologically stressful. Work environments that are isolating, like cubicles, are supposed to increase efficiency, but they actually decrease bonding. In the justice system, prisoners who misbehave are punished with isolation. So in spite of the sense that these methods might make in other ways, they eliminate the opportunity to promote prosocial behavior and attitudes.

"At a fundamental level," Zak says, "my research on trust demonstrates that the basis for human sociality, which requires a 'trustworthiness-discrimination' neural architecture, is love. Oxytocin is the basis for 'mother love.' It passes from mother to infant during breast-feeding. Because humans have an extraordinarily long period of adolescence relative to all other mammals, our parent-child bonding mechanism has to be especially powerful and resilient. It has to extend outward, encouraging us to form temporary attachments to other people, and permitting us to live in large social groups. So it makes sense that oxytocin spurred the human acquisition of intelligence and development of early trading economies. All this from mother love!"

Humans can be pretty perplexing at times, but there's one praiseworthy thing that most of us do routinely. The Investment Game proves that most of us will be helpful to strangers, reflexively, even if we have to spend some resources along the way. Test subjects given 40 IUs of oxytocin—the same dosage I got—are 80 percent more generous about splitting money with a stranger than subjects who got a placebo.

"In reality," Zak says, "we test trustworthiness all the time by studying someone's body language, eye gaze, gestures, and so on. Except for little children and most autistics, we know how to do this automatically. We don't even need to think about it. This is why people like to meet in person whenever there's an important decision to be made, and why we see increased use of videoconferencing."

Another aspect of trust, one that figures into many people's lives now, and which also contradicts prevailing economic theory, is whether an employee who isn't directly supervised can be trusted. Standard theory says that telecommuters will shirk their responsibilities to the maximum extent possible, so they will have plenty of time and energy for selfish pursuits. Their sneakiness would result in what's called "negative utility flow." But telecommuting has become a huge part of many current work lives. In some cases, it can figure into several days every month. In many others it represents almost 100 percent of a job.

If standard economic theory were true, then telecommuting and other business practices where supervision is minimal simply wouldn't survive. But they thrive now, especially since the Internet can deliver an employee's work. The Investment Game suggests that it's an exchange of trust. People trusted with a valuable resource, time, are likely to work hard to reward the company's trust and thereby encourage a positive feedback loop.

It may be possible to trigger your own oxytocin flow. At least, that's one possible interpretation of a study conducted in 2007 by economics professor William Harbaugh of the University of Oregon. Teaming with members of Oregon's psychology department, Harbaugh used fMRI to study how people are affected by the act of contributing to charity.

Using college women as test subjects, the researchers gave each participant $100. Subjects were allowed to keep whatever cash was left over, but first they had to use some of it as a donation to a local food bank. If they didn't contribute, some of their money would be sent there anyway.

In both cases—when the giving was voluntary and when it was mandatory—brain areas related to good feelings were stimulated. Existing economic theory suggests that only the very rich would give charitably, and then when it would enhance their image and hence their business longevity. "But that doesn't happen," Harbaugh says. "There's high participation, where even low-income people are giving away a portion of their income."

Like the participants in the Claremont studies, the subjects apparently were guided by emotionality—being kind and generous simply seemed like the thing to do. Being charitable produced positive feelings.

Read Montague, who did those influential Coke-versus-Pepsi experiments, has been studying trust exchanges for the last four years. His method has been to have two people play a multiple-round trading game while he scans the brains of both players at once, watching how the signals within their brains change as the subjects negotiate back and forth.

Then he got the exceptionally ambitious idea to extend the study across cultures, to make it international. When I last spoke to him, Montague was nearing the end of a two-and-a-half-year experiment in which subjects in Texas interacted with subjects in Hong Kong. With a thirteen-hour difference, plus the complications of working cross-culturally, it was an enormously complex study. However, Montague thinks his team is learning things with big potential impact for international trade, and perhaps beyond that, with vital possibilities for understanding race relations.

The researchers structured the trading game with three variations. Sometimes the players had complete anonymity. No one knew whether the person they were trading with was American or Chinese. In another version, only the Chinese person would know that he was playing with someone from a different culture, while the American wouldn't know the identity of her trading partner. In the third mode, it would be the American who knew that she was trading with someone from a different culture. The idea was to see how cultural signals might produce changes that could be seen in a brain scan, as well as in behavior.

The study is still being written up, but Montague was able to reveal what he called "the punch line." When the Chinese played anonymously—that is, not knowing whether they were matched against an American or another Chinese—they played exactly the same as Americans did when playing anonymously.

Also, their brain scans didn't show any differences at all. But when one player/trader knew the cultural identity of the other, behavior and brain activation changed.

"I don't really know all the subtleties yet," Montague says, "but the bottom line is that it's dramatically different." Apparently, whether we realize it or not, our brains are like computers that automatically switch how they operate when they start interacting with a computer that's running different cultural "software."

The January 2008 issue of *Psychological Science* reported on another subtle study that teases out some variations in cultural software. John Gabrieli, professor of brain and cognitive sciences at MIT and director of the Martinos Imaging Center at the McGovern Institute for Brain Research at MIT, is a tall, calm, unassuming, and brilliant scientist whose research spans the spectrum of the neurosociety. He headed a research team that scanned brain responses in two different groups. Ten subjects were recent immigrants from East Asia; the other ten were Americans. The tests required both groups to make quick perceptual judgments.

Prior research has confirmed that Americans, as members of a culture that values the individual, tend to see objects as things that exist independently from their surroundings. East Asian societies emphasize the collective aspects of life, and members of those cultures tend to perceive objects as existing inside their surroundings. Studies by behavioral psychologists have proven that these different ways of seeing can influence overall perception as well as memory. Gabrieli and his team wanted to know if these cultural differences would result in differing brain activity patterns.

Both groups were given easy visual tests of two kinds. One set of tests emphasized judging the relative length of lines that were positioned near squares, without reference to how big the squares were. The second set involved judging whether some lines were the same size in proportion to the nearby squares. The first task called for an absolute judgment, the kind Americans are usually good at making. The second called for a relative judgment, a type that tends to be easier for East Asians to make.

The brain scans showed much bigger activation differences than the team had expected to find. When the Americans had to make relative judgments, the regions in their brains used for attention-demanding mental tasks were highly stimulated. These regions quieted down when the Americans made absolute judgments. The brain activation patterns of East Asians were the opposite—more activated by making absolute judgments, less activated when they made judgments about the relative sizes of the lines they were shown.

What surprised the researchers, in addition to the pronounced differences in activation, was that whenever a subject's brain attention system had to work outside of his or her cultural "comfort zone," it became engaged in an unusually widespread manner. In a sense, the study produced concrete images of what goes on when a person feels culture shock—that strange sensation you get when you're immersed in a foreign culture and you become dazed by how differently the people think and act in their everyday dealings. But instead of providing an anecdote about one person's experience, or even one group's collective experience, the study produced evidence of a scientific fact. Our brains have to do extra work when they attempt to bridge cultures.

"I think the bigger point," Montague comments, "is that you can study these things now. You don't have to get stuck on pitting your opinion versus mine, fighting to see who is politically more powerful."

That's just one example of how neuroeconomic research into trust has the potential to turn our common future in a better direction. Experiments generated by a handful of brilliant researchers sprinkled across the globe, some of whom you just met in this chapter, are beginning to open the door to possibilities for public policy and personal decisions that could radically accelerate the growth of social capital and consequently make us all happier with our various global and local institutions.

We've known about this neurohormone called oxytocin for only slightly over half a century, and have only recently begun studying how it works in human interactions. Coincidentally, we're at a point in history where we could really use the potential benefit of oxytocin supporting humanity's advanced behaviors to flourish even more. Maybe, as we progress to even greater and more widespread interconnectedness, oxytocin research will help teach us how to make the necessary transitions in a more easeful, trust-based, natural-feeling way. For example, the studies by Montague and Gabrieli could be the beginning of a fundamental reshaping of policies and procedures among diplomats, businesspeople, teachers, and all others who find themselves working in cultures that are very unlike their own.

Trust-based research will ultimately broaden the science of ergonomics. We will go beyond just human-body-friendly design in furniture, appliances, and tools to create human-brain-friendly design in the institutions that affect our mental and social lives.

From an economic point of view, if an oxytocin test were inexpensive and readily available, it would reduce cheating. That alone would raise living standards. Think about the resources we burn enforcing contracts, at the national and state levels, and as individuals personally. This is a huge economic loss, and it could be avoided by accurate scientific measurement of trustworthiness.

As we learn more about the neurobiology of trust—how people make decisions, and all of the components of human social interaction that go into decision-making situations from contract development to dispute resolution—the transaction cost of doing business will drop astoundingly. We will be able to reorganize business units, corporations, and entire economies around a stronger basis of trust. The neurosociety will be characterized by flatter organizations, with less hierarchy and more heterarchy. Currently, CEO salaries can be more than five hundred times higher than those of average workers. In the neurosociety, corporate wealth will flow in a more lateral way, decreasing the gap between the haves and have-nots, bolstering the middle class, and reducing poverty. That development will add to our social capital, making prosperity last longer.

Neurotechnology will also provide new tools for management. It will become less seat-of-the-pants and start being something of a science. Today's best managers are masters at understanding human motivation. They tailor their tone, attitude, strategies, and tactics to the needs of each individual situation. But many people who get to be managers are often the fiercest competitors, and they don't always have a good empathic skill set. In the future, more people will have better tools—in training, perhaps in neurofeedback, even in exquisitely targeted neuropharmaceuticals—to help them have the same kind of managerial mastery that now comes naturally to the gifted few. Management styles will be about enhancing individual trust, emotional stability, and cognitive clarity. In turn, we will see improvements in team performance and the social pleasures of working toward common goals.

Imagine some of the other things that could happen if neuroeconomic research can deliver on some of this oxytocin-promoting promise. Connectedness could be a bigger part of our day-to-day lives. Students might feel emotional as well as practical reasons for building their potential. They will care more about their schools and the friendships nurtured there, and will want to stay in rather than drop out. Workers will feel there's a direct connection between prosperity for their employers and better lives for themselves, and employers will gladly devote resources to keeping that connectedness well rewarded. Families will be more about bonding, less about bickering.

All because the oxytocin-induced ease I experienced in the Claremont laboratory will no longer be so rare.

SIX

DO YOU SEE WHAT I HEAR?

Art is nearer to life than any fact can ever be. —Ananda K. Kumaraswamy

The capacity to be puzzled is the premise of all creation, be it in art or in science.
—Erich Fromm

In today's teach-to-the-test educational environment, art is a brain-candy option, an afterthought, a pretty park situated somewhere outside life's more vital concerns. You might visit art of some kind now and then, seeking some rainy-day fun for boys and girls.

Up in our complicated craniums, art is a whole lot more. It's humanity's first information superhighway. It's our private means of dealing with the full range of everything we are able to individually think and feel, and our social means for sharing all those thoughts and feelings. Art transmits extremely nuanced ideas about what it means—at any given moment in history—to be a woman, a man, a girl, a boy, a participant in life, a voyager bound for death and possibly beyond, or for exploring any other concept that leaves its traces while passing through the nearly infinite circuitry of our brains. It's a close kin to religion, one who doesn't take herself quite so seriously—but who still knows that she matters plenty (this random gender assignment comes from the Greek idea of the Muses, who were female).

Since art is undeniably one of the most complex and meaningful activities our brains have ever invented, it's also a mother lode of rich questions to ask about how and why our minds do what they do.

Jonah Lehrer examined the lives and work of eight creative giants (Whitman, Eliot, Escoffier, Proust, Cézanne, Stravinsky, Stein, Woolf) in *Proust Was a Neuroscientist*, his 2007 book based on the concept that each of these artists independently figured out something vital about human brains and sensory perception, and that science is now proving their insights to be true. However, it's also important to know that neuroscientists are happy to take cues from the imaginative powers of artists. They aren't threatened by the fact they're playing catch-up with the various ideas about the nature of the mind artists have toyed with for millennia. They're emboldened by the clues artists have laid out, and thrilled by the possibility of building on knowledge that first arrived because an artist's brain expressed it intuitively.

Just as our buildings are shaped by the available resources of construction technology, artworks are shaped by the functional abilities of our brains. Both in creating art and enjoying it, our experience is defined by the capacities of our brains. A new discipline called neuroesthetics uses art as a pathway into understanding the organizational workings of our brains—in particular how our senses relay their messages into the brain, and what happens with that sensory information as it gets combined, restructured, and assimilated.

So far as we know, every human culture has made art. This is a curious fact. Art isn't necessary for the survival of our species—at least not on an obvious level. Art won't quell physical hunger, or keep bitter weather from freezing our bodies. Art didn't keep terrifying wolves or saber-toothed tigers away from our ancestors' caves. So why does every human society spawn art and artists? How does something so unnecessary generate billions in commerce yearly? Why are the most influential art and artists enshrined by both common and highborn people? Why are they feared by autocrats? (When some especially powerful West African griot singers died, tribal chieftains would place their bodies in hollow trees far from the village, fearing that their powers would linger beyond their lifetimes.)

One way or another, artists have been stirring up cultural, personal, and political change in our brains for as long as they've been mixing pigments, twanging tones, clacking dry bones, pounding drums, snapping pics, rapping or writing rhymes, and so on.

Philosophers have worked over the centuries, trying to define the nature and the importance of art. Immanuel Kant's *The Critique of Judgment* explores the workings of the mind through art. Large swaths of Schopenhauer's *The World as Will and Representation* are concerned with art. He defined "genius" as a trait all of us have, though some more than others, and he believed that our ability to receive esthetic experiences is a sign of genius.

Before brain imaging technologies reached their current level of development, ideas about art were as unprovable as they were intriguing. Now students of neuroesthetics can dream up provocative experiments to test theories about why art affects us. They have begun peeling away the theories that don't work, and are going deeper with the ones that seem to yield pay dirt.

As work in neuroesthetics accelerates, the combination of artistic expression and brain scanning will give us more solid and provable insights about humanity. The sensory satisfactions of fine chocolate, the complexities inside handcrafted wines, the excitement of bold architecture, music that—in Shakespeare's phrase—"can hale men's souls out of their bodies," paintings that compel us to gaze again and again, dances that seem to impel the very turning of the planet, stories that rivet both readers and listeners: All of these forces can now be shown as brain activations on the computer screens of scanners.

Synesthesia is a terrific starting point for understanding how science and art now intersect. It's the brain condition that blurs the boundaries we traditionally use to divide the various territories within the realm of our senses. Synesthesia, which is passed along genetically, leads beautifully into many more facets of neuroesthetics, and opens up first glimpses of some of the truly amazing ideas now being pursued.

Synesthesia is the condition—though most who have it consider it a blessing—that occurs if someone receiving a sensory message also gets simultaneous impressions in other sensory pathways that seem unrelated. For example, hearing music may awaken real-seeming sensations of colors, images, or flavors, or of some other nonaural sense experience. We call someone with this condition (or blessing) a synesthete.

The first scientist to describe synesthesia was Francis Galton, a younger cousin of Charles Darwin who made great contributions to statistical analysis in science, and to the study of human mental capacity, psychometrics, a field he created. It was Galton who discovered synesthesia runs in families.

In recent research, Simon Baron Cohen has verified Galton's findings. (Meanwhile, Simon's nonscientist brother, Sacha Baron Cohen, has proven that people can become gullible around an actor pretending to be a Kazakhstani journalist named Borat or a goofball rapper named Ali G. It's fascinating that one brother tests mental capacity while the other tests the limits of taste and decorum, both siblings doing memorable work.)

Some synesthetes have just a small trace of these multiple sensed responses, perhaps an emotional response to certain letters when they appear at the start of names. Others experience more full-blown synesthesia, with sounds, forms, feelings, and colors doing the tango to various sensual perceptions throughout the day.

There's a certain variety of synesthesia called "mirror-touch." Michael Banissy and Jamie Ward of the University College London published a paper in 2007 detailing how some observers will get activation in their brains, in the same areas activated by touch, simply from seeing someone else being physically touched.[1] This just might be the neural structure that underlies compassion, an evolved and crucial human trait, and the tendency to care about the welfare of others, to figuratively and perhaps even a little bit literally *feel* what they feel. The researchers found that people who tested high for mirror-touch synesthesia also scored high on a questionnaire designed to measure empathy.

Wassily Kandinsky, the founder of abstract art, didn't need an iPod or CD player set on Shuffle in order to fill his ears with music when he painted. The colors on his palette evoked his own personal vibrant music. Kandinsky was born in Moscow in 1866, and could play both cello and piano at an early age. Painting let him formalize the relationships he perceived between colors and sounds. "Color is the keyboard," he once explained. "The eyes are the harmonies, the soul is the piano with many strings. The artist is the hand that plays, touching one key or another, to cause vibrations in the soul." Yellow evoked the note C for him, in the timbre of a brass instrument. ("Timbre" refers to the vocal character of an instrument or singer. It makes a C played on a violin sound different from one coming out of a tuba, even though it is technically the same note.)

Kandinsky painted his last great work, *Composition X*, during the turmoil of

World War II. He chose a black background because that color sounded, for him, a note of finality. He heard combinations of colors as if they were chords, more dissonant in some combinations, sweeter and more consonant in others. Shapes also triggered associations, with the circle summoning thoughts of peace.

Another artistic giant who came from Russia, the novelist Vladimir Nabokov, was also a synesthete. Mind-stretching clues abound in his writing, including a character who compares the word "loyalty" to a golden fork lying in the sun, and another who hears "the distinct sound of real orange-hued mind music." Nabokov's synesthesia linked numbers with colors. He described these unusual sensations in his memoir, *Speak, Memory*. His wife, Vera, had the same type of synesthesia, but saw different colors. Their son, Dmitri, experienced numbers in a way that blended the colors that his parents perceived, a fact that underlines the genetically transferred nature of synesthesia.

In midsummer of 2007 I had the pleasure of walking the streets of Boston with Marcia Smilack, an artistic photographer with a doctorate in English literature from Brown University. She experiences a multilayered synesthesia, and the photos with which she records her unique perceptions have been widely exhibited. Smilack seems to have almost every possible form of synesthesia, except for the relatively common one in which colors are associated with numbers. When she looks at topographical maps she sees time. She visualizes concepts as shapes. A year, for example, is an oval.

Before telling me this, she posed an interesting question: What shape did I think a year might possess? I reflected for a moment, then said an oval. Does this mean that I have a very mild form of synesthesia? Maybe. A neuroscientist would say that the importance is not actually the shape I picked, but the strength with which the connection between the concept and shape appeared in my mind, and whether the connection is reproducible.

We happened to pass, as is typical on the streets of Boston—home of the Berklee College of Music—musicians performing for tips on the sidewalk. The first encounter was with a jazz duo featuring a young woman who played the violin. I don't know enough about music to criticize her playing; I can only say that I didn't find it pleasing. Smilack agreed and saw lines of color in the air, emanating from the woman and her instrument. The lines were jagged, making abrupt turns that didn't seem consciously designed.

A few blocks later, we passed a young woman playing her saxophone to the open air. I liked how she sounded. So did my companion, but she also saw curving lines looping around and around the saxophonist, visual expressions of the beautifully realized design heard in the performance.

When Smilack was a little girl, she reached up and plunked a single note on the family's piano. The note was green. She supposed that everyone else got color with their music too. Twenty-five years later, in 1979, she was at a laundry and, thinking no one else was there, enjoyed herself by dancing to the rhythmic pulses of the dryer. Then she saw another woman, a psychology grad student as it turned out, who had been watching her spontaneous dance.

They conversed for a short time about music and art. Smilack had never before told anyone about her multisensory perceptions, but at the end of their talk the woman remarked, "I believe you may be synesthetic." Intrigued, Smilack pursued the topic just a short time but eventually let it drop.

Twenty years later, in 1999, she read a *New York Times* article about Carol Steen, a New York City artist who is also a synesthete. Steen's words expressed what Smilack had experienced herself, but had never fully articulated. She realized that she had been using synesthesia in her artwork for years, intuitively, without realizing she was tapping into something extraordinary. She fired off an email to Steen, saying, "I hear with my eyes."

Steen answered immediately, typing, "Welcome to the club; you're in great company."

Smilack has evolved an artistic process around her synesthesia. She goes somewhere that seems interesting, then looks and listens until she notices the arrival of a synesthetic reaction. It might come in the form of motion, taste, a sense of texture, or some combination of sensations. When it comes, she takes photographs of the scene that evoked her response. She is drawn to reflective water surfaces, and waits until the reflection picks up movement from the wind. As she describes it, the sea is her canvas, the wind is the brush, and the season and the location provide coloration. Buildings often appear, their shapes intriguingly rippled.

Smilack trusts that intuitive signals will arrive faster and more reliably than formed thoughts. Rather than consciously picking a subject and deciding what angle, composition, and light will create the most effective photo, she tries her best to *not* think while waiting for the signals from her body. She avoids looking directly at the scene in front of her lens. Her intent is to show, and to remember always, that beauty lurks all around us, "inside time and space and all the stuff in the middle, including consciousness," waiting to be perceived.

V. S. Ramachandran, a prominent researcher at the University of California, San Diego, believes that the sensuous dance of synesthesia may teach us a lot about how language evolved, as well as the human capacity for abstract thought.

Looking into how synesthesia happens in the brain presents a basketful of clues about the art-loving aspects of our brains, and helps explain why we treasure music, food, plays, movies, and the many other forms of our panhuman activity called art, which the Balinese sweetly refer to as "bringing the gods down to Earth."

Most of us easily understand a phrase like "People who live in glass houses should not throw stones" (although a stroke or other brain damage can sometimes block our ability to decipher figurative language). Almost everybody agrees that certain colors, critical remarks, aged cheeses, or Hawaiian shirts—none of which could actually cut wet paper—can be sharp. Music is sometimes cool. Exciting ideas and sexy bodies are hot.

This shared way of thinking, our ability to find meaning in nonsensical language, is a clue. It suggests that we may all have at least some little bit of synesthesia, possibly a smaller-scale version of the same brain conditions that produce true, full-blown synesthesia.

I spent months pursuing an interview with Ramachandran. (He may have become a little media-wary since some of his findings—as you'll see in the next chapter, "Where Is God?"—have gotten overhyped and misinterpreted in the popular press.) Ramachandran believes that about one person in twenty possesses some degree of synesthesia, but that it is seven times more common in artists, poets, and novelists. "And," he asks, "why should this be so?"

Some people have proposed that number/color synesthesia is simply a persistent memory, possibly from an association made from a favorite book in childhood. Ramachandran and his team looked instead at the idea that synesthesia occurs because synesthetes have cross-wiring between brain areas that are key to sensual perception. Testing synesthetes, they found that when they changed the size or shape of a letter, the subject would see a different color than he or she saw before. So it wasn't memory. It was an unusual, maverick sort of activation in the brain.

Since then, other researchers have actually been able to show that synesthetes have an increased amount of nerve fibers between brain regions. "So," Ramachandran says, "it's about as good as it gets as far as confirming a theory." He and his colleagues have tracked some kinds of synesthesia to a brain structure that is in the temporal lobes. In it, a color-perceiving area sits right next to an important area for visually recognizing numbers, almost touching. Color-with-number synesthetes have some extra connectivity in these neighboring brain parts.

Ramachandran's theory is that synesthesia happens because of a pruning gene. In a fetus, everything is wired to everything else. Normally, during fetal development or later in childhood, pruning genes come in and trim the excess connections, creating the separate modules that we find doing different tasks in adult humans. If there's some mutation in that gene and the pruning doesn't work completely, there will be cross-talk. If this cross-talk is extensive enough, the result is synesthesia.

"Now," he continues, "you have to explain why it would be more common in poets, artists, and novelists. And the answer is this: If the mutation happens diffusely throughout the brain, you're going to get greater cross-connections. Since different ideas and concepts are represented in different regions of the brain, what if metaphor—which is the great thing that poets, artists, and novelists have in common—comes from linking seemingly unrelated words and ideas? The mutation would give you greater opportunities to form metaphors."

To build his example, Ramachandran suggests that every word represents a cluster of associations, a penumbra. When Shakespeare wrote, "Juliet is the sun," he wanted us to feel how the penumbras of association for two different things actually overlap. Juliet is a woman. The sun is not. Juliet lives in Verona. The sun is millions of miles away. But Juliet is warm and the sun is warm. Juliet is nurturing and the sun is nurturing. Juliet is radiant and the sun is radiant.

"This overlap of two penumbras," Ramachandran continues, "is the basis of metaphor: We extract what is common between Juliet and the sun. If connections in one's brain leave the penumbras more closely connected, then there's a greater overlap and greater opportunity for metaphor, making people artistic. That's what we think caused the persistence of this gene. Why else would one out of twenty people have synesthesia? If it was a useless gene, it would get weeded out by genetic drift. But that didn't happen. I think it's because of the hidden agendas of the gene. It makes some people more creative."

"Which," I venture, "has a selective advantage?"

"Yes," he says. "This is why you constantly hear artists talking about synesthesia. The old idea was that they were all nuts. Now they can see that their differences from normal people are valuable."

Neuroesthetics researchers tend to look at basic early physiological mechanisms as a clue to understanding why we respond to art. Ramachandran looks at higher cognitive processes as well. For example, certain parts of the brain activate when we look at faces, but they will respond more intensely for artistically created faces than for average, everyday faces. "Why is a face from great art more evocative than any old face?" Ramachandran asks. He believes that three key things are at work when we feel an esthetic reaction.

The first is the anatomy of the brain, the physical hardware of our mental computers. Second is the set of psychological laws behind our responses—like whether we get a pleasant "aha" jolt on seeing groups of related things brought together, for example. Most of these psychological laws remain to be discovered, but with brain imaging, scientists can make specific predictions about psychological laws and actually test them, a capability that has been available only over the past decade. "So you can go on a fishing expedition," Ramachandran says. "That's what you first do in science. Then you find there are some patterns there, perceptual laws or esthetic laws, and you can define what they are."

The third aspect is evolution. Why did these laws evolve? How did they help earlier humans stay alive, and reproduce? Here's a surprising clue. In a widely quoted 1999 paper Ramachandran and his colleague William Hirstein predicted that brains would respond even more strongly to caricatures, drawings such as political cartoons, which hugely exaggerate the features of a face.[2] Recent research has confirmed that idea. Ramachandran believes the reason for this quirk of human perception can be found in a book called *The Herring Gull's World*, written by the Dutch scientist Nikolaas Tinbergen, winner of a 1973 Nobel Prize.

"Why," Ramachandran asks, "would you respond to a distorted caricature? It can be explained in terms of Tinbergen's gull chick example. He noticed that baby chick herring gulls, when they want food, start pecking the beak of the mother. The mother then regurgitates food.

"The mother's beak is a long yellow thing with a red spot on the end. Tinbergen found that you could just wave a beak around and the chick would beg food from it. So as far as the chick is concerned, the beak equals the mother. It gives him a simpler computation, a shortcut to recognizing his food source. But then Tinbergen found out that he didn't even need a beak. If he took a long yellow stick and painted it with three red stripes, the chicks would respond even more. It has to do with a kind of code that is in the chick's brain, concerning the mother's beak. If you create a kind of superbeak, you excite the baby bird more than you can with the real beak.

"I think that when we respond to abstract art, or semiabstract art, we're behaving like the seagull chicks. We're seeing a stylized, exaggerated version of some kind of pattern that gives us an "aha" of pleasure, of arousal. This seems to happen at multiple levels with great works of art, and these levels kind of harmonize or resonate with each other, creating a multiple, final, climactic aha of a great work of art."

I always wonder how any new research might end up being applied to daily life, so I ask Ramachandran if he thinks artists will begin using this kind of knowledge to help them make more evocative art.

"I think so," he says. "A lot of artists, when they learn about the research, say, 'Wow. Now I know I'm not crazy! I'm actually using these laws.' As they become more conscious of what they're doing, they can start doing it even more."

"To take it a step further," I ask, "do you imagine the knowledge being applied in trying to develop new forms of art?"

"Very much so. I think that an extremely talented artist could take advantage of the principles and actually deploy them. There's a guy named Bruce Gooch at Northwestern, and he has developed computer algorithms that mimic some of the laws."

When I checked Gooch's Web site, I discovered that he is doing a lot of far-out work that extends from his 2003 Ph.D. dissertation, entitled "Human Facial Illustrations: Creation and Evaluation Using Behavioral Studies and fMRI." Looking at his images and publications, I recalled Ramachandran saying, "So in theory, in the distant future you may have a computer generating pleasing works of abstract art using these principles."

Language is a medium of art, of course, and so is music. Brain scanning has shown that both of them use several regions in the brain, and that they also share many regions. Charles Darwin believed our ancestors had a good grasp of music before they evolved language. He thought that music may have helped our ancestors find mates, allowing them to stand out from competitors by making memorable sounds, just like a songwriter of today tries to get lyric and melodic "hooks" into his or her composition, making it hard to forget. Presumably, when a man or woman returned to the tribal den after a day spent tracking food, a potential mate would remember that the returning person matched up with a set of pleasing sounds, and would be happy to see him or her—maybe happy enough to roll the genetic dice together, ensuring another generation of music lovers.

Nearly everyone acknowledges the connectedness of music and emotion. But until brain imaging arrived, very few in academia wanted to probe that confluence. Emotion has been suspected of being the opposite of rationality, something to be overcome instead of valued.

Good management of emotions is a skill package that takes lots of work to acquire, but emotions themselves can be seen as our crucial early-warning systems, important response mechanisms for coping with change. As the artistic process Smilack has evolved shows, impulses have the advantage of speed. They arrive faster than we can combine sense impressions into coherent whole structures. When immediate decisions are needed, emotions give us a fast response.

Steven Mithen's 2005 book, *The Singing Neanderthals: The Origins of Music, Language, Mind, and Body*, proposes that language and music evolved more or less at the same time. The fact they are both found in multiple areas of the brain indicates both are evolutionarily important, to a high degree. It's possible to lose function in certain parts of one's brain, yet still hold on to key aspects of language and music. Language and music cause our brains to coordinate in several different regions. This is true whether we're creating with music or language, or just listening.

Daniel J. Levitin pointed out in his 2007 book, *This Is Your Brain on Music: The Science of a Human Obsession*, that victims of Alzheimer's disease can lose massive amounts of memory and still remember songs, especially the songs of their younger years. According to Levitin, the emotional centers of the brain work together with neurotransmitters to "tag" our memories of the music we find to be emotionally charged. That explains why each generation grows nostalgic about its music, and why "oldies but goodies" radio formats are easy to find on the dial. Teenage years are emotionally charged—if not to say volatile—and connections to that music are etched deeply. Levitin is a former rock-and-roll guitar player who came to neuroesthetics by a wandering road. Years ago, his band got signed, went into a San Francisco recording studio, but then imploded. Fortunately, the sound engineers noticed his fascination with their work and taught him enough to get him started professionally in their field. Later on, courses taken at Stanford motivated him to go all the way to a doctorate. He now teaches at McGill University, specializing in neuroscience and music.

McGill is a hot spot for music and neuroscience these days, along with the University of Montreal, especially since the 2007 opening in Montreal of a $14 million center called BRAMS. The five letters stand for BRAin, Music, and Sound research center. Prime movers in the center's conception include Robert Zatorre, a McGill neuroscientist, and Isabelle Peretz, a University of Montreal psychologist.

BRAMS includes a concert hall in which scientists can study how listeners react to music, and a soundproof studio outfitted with a special Bösendorfer piano that is wired into a computer and flanked by twenty-four cameras dedicated to recording every possible physical nuance in the performances of pianists.

Oliver Sacks made music and neuroscience the focus of his newest book, *Musicophilia*. Sacks cites some of Zatorre's neuroimaging research, in which subjects who imagined hearing music showed activation of a major hearing area in their brains, the auditory cortex, almost as powerful as the activation in subjects who actually heard the music performed.

Sacks explains that the activities of hearing, composing, and performing music activate that auditory cortex, plus the motor cortex, plus other brain regions involved in choosing and planning. The fact that these regions can kick into musical gear even in the absence of sound explains one of the most amazing facts in the history of art—Beethoven could still compose music even after becoming deaf.

It's not just the combining of sounds and motor skills and planning that makes for memorable music. Another key ingredient is a talent for combining musical elements in ways which are unexpected, yet fully satisfying.

In my hometown of San Francisco there lives a remarkable singer-songwriter named Jesse De Natale. He invents some sound and imagery combinations I enjoy, but could never create myself. For example, "Nightingale" is a song that's about death yet is also bouncy and somehow cheerful. The song tells listeners to focus on lovingness because everyone we know eventually will be "rainbow gone." We typically think of dead people as being in the ground, in a coffin. De Natale seems to see them as radiant, somewhere up in the sky, possibly visible to us just once in a while. It took a wonderfully untypical brain to articulate that thought for the rest of us, and to place it inside a well-tuned melody.

According to the highly regarded British researcher Semir Zeki, creativity lies in seeing relations other people have not seen before. He believes, much as Ramachandran does, that artistic and other creative people have some sort of in-brain connections the rest of us don't have, but which we can possibly develop to some degree. If that notion is right, we don't have to just admire creativity; we can also learn how to get ourselves enjoyably into creative modes. After all, as musicians say, you don't "work" a musical instrument. You *play* it.

Zeki is a synesthete. The letters in a word can stir his emotions in various ways. Specifically, he senses a personality in a word, inspired mostly by its first letter. When the city once spelled Calcutta got modified to Kolkata, it began to evoke a very different feeling in his mind, one less pleasant.

Zeki lives in London and has been fascinated throughout his life by opera, art, and literature. He is also one of the world's most distinguished visual neurobiologists, a pioneer in neuroesthetic research, and a professor at University College London. Every January for the past several years I've met with him in Berkeley, where we both attend the annual Neuroesthetics Conference hosted by the Minerva Foundation. Zeki, in fact, won the very first Golden Brain Award the Minerva Foundation ever granted, back in 1985. It's sort of the neuroesthetics Oscar.

Conversations with Zeki are always intense, revelatory, and immensely fun. "Synesthesia was for a long time regarded as an aberration," he recently told me. "And indeed it is. But I would be extremely disappointed if that aberration were ever taken away from me. It has enriched my life very significantly."

Excited, though in a thoroughly British and politely modulated manner, Zeki revealed he had just secured the first humanities-related grant ever to be given by the Wellcome Trust, the world's largest medical charity. (Crossing traditional boundaries in universities, hospitals, and other institutions is a hallmark of the Neuro Revolution.)

"I think what's going to happen in the coming three or four years," Zeki told me, "is the realization that there are *common* problems that different disciplines have simply been addressing in *different* ways."

For example, he noted, even though one's first thought about esthetic experiences is probably linked to pleasure, these experiences can also evoke pain. "The pleasure and reward centers of the brain are being studied in great detail, but it's also known that a great work of art, such as Michelangelo's *Pietà*, is profoundly moving yet also painful to contemplate."

Michelangelo created many sculptures on the theme of Jesus brought down from his crucifixion, limp and lifeless in the arms of his mother, Mary. The most famous *Pietà* is the one on display in St. Peter's Basilica in Vatican City. It presents an image in pale marble of the deepest grief and unimaginable loss. Yet this statue reflects a kind of serenity, perhaps in the composure of Mary's face. She seems to know her son's resurrection will eventually follow.

"How can we enjoy a work of art that is painful to see? This question has been asked by philosophers for at least two thousand years," Zeki said. But it has not, until very recently, been considered an appropriate question for a neurobiologist, for the "very, very silly reason that such questions belong to the humanities. But of course," Zeki continued, "they don't. These questions belong to science. These questions belong to inquisitive minds."

At an earlier point in Zeki's career he stopped and wondered: Why study the visual brain in such detail if, as a scientist, he could not utter a single word about what happens to our brains when we look at something that we find beautiful? In 1994 he was asked to give the prestigious Woodhull Lecture to the Royal Institution of Great Britain. He took a chance and asked the director if he might speak on art and the brain. "I thought he would slap me down," Zeki recalled, "and tell me, 'Come on. This is a scientific institution!' " Instead, Zeki received an enthusiastic approval, which encouraged him to keep on pursuing his esthetic interests scientifically.

In recent years, and all the more decisively since the advent of fMRI, neuroscience has established a place within almost every single field, including chemistry, pharmacology, physiology, computational science, and anatomy. The humanities are the final frontier.

Zeki asserts that the humanities and neuroscience really *belong together*, and each has the power to stimulate growth in the other.

"There are problems proposed by the humanities, or by philosophy or by art," he said, "which are of enormous interest to neuroscience." For example, the French painter Cézanne liked to portray both the landscape and the homespun people of Aix-en-Provence, in the rural south of France. He often returned to the same subject, such as a nearby mountain peak, but painted it in subtly different ways each time. Cézanne is closely linked with impressionists like Pissarro, his great teacher. But his art gradually evolved in directions that pointed toward the cubism that Picasso and others expressed.

"Cézanne's preoccupation," Zeki told me, "was to try to see how form is modulated by color. Now, we know that form and color are mapped by separate systems in the brain. One of the problems of visual neurobiology is to ask how the two separate systems interact to give us apparently unified perceptions, with the colors reattached to the forms."

Kinetic art, whose examples have the property of movability, also fascinates Zeki both as an art lover and as a scientist.

Marcel Duchamp constructed the first piece of kinetic art, in 1913. Called *Bicycle*, it is simply a bicycle's front wheel mounted in a conventional set of front forks and fastened upside down to the seat of a four-legged stool painted white. The most famous kinetic artist was the American sculptor Alexander Calder. He moved to Paris in his late twenties, always carrying a spool of wire and pliers when he went to parties. He would twist the wire into whimsical forms, toys, and circus-themed objects. Calder invented the mobile, a suspended sculpture with parts that move independently.

As kinetic art developed, form and color became less important to the works, and motion more important. "The emphasis was really on moving objects," Zeki told me. "Things which made no sense." Kinetic art was loved by the avant-garde, but most people didn't get it.

Today almost every kid in America grows up watching a mobile sculpture twirl in space over his or her crib. Calder's invention is thought to help stimulate brain development and lull children into sleep.

According to Zeki, there's a neurobiological fact that may explain why kinetic art fascinates babies as well as grown-ups. We all have a brain region specialized for visual motion but not interested in form or color.

Neuroscience, Zeki believes, is "if not the queen of sciences for the coming century, certainly likely to be one of the princesses. No doubt of that. People are very interested in the brain and its workings. But they may find it hard to follow strictly scientific terminology. They relate much more readily to something such as, 'Let me tell you today about love and beauty—the neurobiology of love and beauty.' There is no more fascinating subject than that. So our love of beauty is going to be a vehicle for acquainting people with a scientific method and a scientific discipline which is going to be of enormous importance in the future."

Zeki believes we will see neuroesthetics applied in practical ways in the near future. For instance, there's that irresistible but often tricky enterprise our Founding Fathers called "the pursuit of happiness." Most of us would like having more happiness in our lives; we're just not always sure what will deliver it. We're trained, by schools and religion, parents and other pervasive influences, that pleasure and sin walk hand in hand. Which may be true, sometimes, but which also can be a source of confusion and misdirection. It can lead to what's been called the Listerine Theory of Life: If anything feels good *to* you, it can't be very good *for* you.

"At what point," Zeki asks, "can a person, in neurobiological terms, be considered to be a happy and satisfied person?"

That's the central question of Sigmund Freud's 1930 book, *Das Unbehagen in der Kultur*, which translates as "The Unease in Culture"; English-speaking readers know it as *Civilization and Its Discontents* (a title shamelessly parodied by the Fibonaccis' 1987 art-rock album, *Civilization and its Discotheques*).

Freud noted that people, in spite of all the extraordinary and still-accumulating achievements of Western civilization, are generally not very satisfied. So what do they seek? They seek happiness. But what does seeking happiness amount to? Freud called it "the satisfaction of the Pleasure Principle." This valuable yet vague insight is something neurobiology might help us pin down. For example, we may develop a sort of relative emotional gauge to help individuals know which are the conditions that most satisfy them. It's not entirely in the realm of imagination.

Just days before we met, Zeki had sent a manuscript entitled "Splendors and Miseries of the Brain" to an academic publisher. The title is adapted from an evocative Gustave Flaubert novel, *The Splendors and Miseries of Courtesans*. Zeki's book sifts rigorously through ideas about the future of neuroesthetics. He foresees in perhaps twenty years, people reading Flaubert's *Madame Bovary* or Dostoevsky's *The Brothers Karamazov* might come across a poignant bit of description and realize how closely it fits with neurobiological facts that have by then become common knowledge.

Neuroimaging will also help us pin down exactly how and why a work of real genius affects us differently than more pedestrian artwork. "Much of great music is improvised," Zeki told me. "There's also a lot of learning behind it. But the great improvisers, like John Coltrane and Ray Charles, achieved some magical effects. I listen to them again, and again, and again. They hit this just-right note, and it seems to remain fascinating and perfect forever.

"Now, it would be extremely interesting to look in the scanner while people hear notes produced with such perfection, and then listen to other versions of the same material which are not in the same league. Where in the brain would the activation be different? Can we arrive at any objective measure of the satisfaction I derive, and others derive? Because there are millions of listeners who never tire of Coltrane and Louis Armstrong and Ella Fitzgerald. I've been listening to Ella Fitzgerald for years, and I just cannot detect a single wrong note from her. The inflection of her voice at certain points, it's a work of genius. But it's improvised."

Putting people into scanners to understand such fine points is something neuroesthetic researchers will do much more often in the coming years. Looking far into the neurosociety, it is possible to envision a whole series of developments that will emerge from this research to transform how we experience the world around us.

As neurotechnology advances, we will have the ability to work backward, to reverse-engineer the experiences generated by art and use some future variant of fMRI to tell us how to construct great songs, plays, and paintings. A day will arrive when, for some, becoming a great artist will mean learning the fundamentals of neuroesthetics. This means that new forms and styles of artistic creativity will emerge that have not been conceived of yet. Perhaps even something as new as kinetic art lies in the shadows of our neurons. Moreover, a host of new tools and technologies will alter how we experience art and entertainment, going beyond the emotion-reading neurotainment video game systems you've already heard about in chapter 3.

As we peel away the underlying neurobiology of pain, addiction, and pleasure, a new form of experiential enablers will emerge in the form of pleasureceuticals. These nonaddictive, pleasure-inducing neuroceuticals, which may come in the form of a small drink, will eventually augment and then replace the addictive and destructive choices of the preneuroscientific era. These new tools will have highly refined capabilities to induce, for example, specific synesthetic experiences for a defined period of time or even empathy perfectly timed for the compassionate moment of the concerto.

Neurotechnology of all types, including neurodevices that will noninvasively stimulate different regions of the brain to lightly induce sensations, will be used in concert with full virtual reality environments to bring about entirely unique experiential entertainment landscapes made possible with the help of neurotechnology. Seem far-fetched? Just look at history, wherein new technological tools made entirely new forms of art flourish. For example, electricity brought about the advent of the cinema (first without sound and then, incredibly, with sound), and the microchip has made possible digital animation effects that are the backbone of today's video game entertainment systems or the electronic music closely listened to by hundreds of millions of people across the planet.

Of course, much stands in the way of the development of these new tools, not the least of which is a framework for global drug policies set by three UN conventions, dating from 1961, 1981, and 1988. Between them, these conventions set rules prohibiting, in almost any circumstances, "the production, manufacture, trade, use or possession of potentially harmful plant-based and synthetic non-medical drugs, other than tobacco and alcohol." Interestingly, these conventions don't mention anything about noninvasive external stimulation systems.

Recently, several countries, including Australia and Canada, have begun to question the logic of this global prohibition, and are considering legalizing certain illicit drugs while placing a heavy emphasis on "harm-reduction" programs. While it is good to see that governments are asking some of the right questions, the legalization of harmful and addictive substances is absolutely the wrong answer. The knowledge is in reach to develop new, nonaddictive recreational tools that can induce an immense variety of specific pleasant sensory experiences. The development of these nonaddictive tools will reach far beyond expanding artistic expression, impacting courtrooms, classrooms, and boardrooms across the planet.

The realm of the senses is also the realm of delight. The Neuro Revolution is going to let us travel further, with greater satisfaction, into this infinite territory.

SEVEN

WHERE IS GOD?

I worry where tonight fits in the Cosmic Scheme of things.
I worry there is no Cosmic Scheme of things.

—Lily Tomlin

To spend more time in learning is better than to spend more time in praying.

—Muhammad

Carefully—for this is the world's most dangerous territory—but resolutely—because science is fairly proud of what it has done to advance humankind so far—neurotechnology has begun stepping into the sacred.

These steps are being taken thoughtfully and methodically, with healthy respect for the profound depth religion holds in the lives of believers, and with solid awareness that in the history of human thought there's no trickier minefield than religion. As Lily Tomlin once commented, "When you talk to God, that's called praying. But when God talks to you, that's called madness."

Fully half of Americans say they have experienced a life-altering spiritual event they consider a turning point in their lives. But according to the National Opinion Research Center, nearly one in five Americans report experiences that could be listed on the pages of the DSM-IV, the primary diagnostic manual for describing mental illnesses, such as hearing God talking to them, floating outside their bodies, or being contacted by the dead.

Even though Nietszche's most quoted line is "God is dead," a great many researchers of various religious and nonreligious backgrounds now believe that what we call the divine lives in our neural circuitry, and that science can now both evoke its presence and also help us understand it more fully. Consciousness, and in particular its connection to the infinite, will always hold mysteries we cannot penetrate. But consciousness is evident in regions of the brain, and now pictures can be taken while it is at work. This combination of brain imaging and the study of religion has made possible neurotheology, a field in which science studies the brain while the brain looks toward the source of all creation.

The first book on the field was *Neurotheology: Virtual Religion in the 21st Century*, published in 1994 by Laurence O. McKinney. It's one of at least nine neurotheology books in recent years. McKinney directs the American Institute for Mindfulness in Arlington, Massachusetts, and he's one of the founders of *New Age* magazine. He believes there doesn't have to be a conflict between theology and technology. As he notes concerning Western Reformed Buddhism, "Relying on modern neuroscience rather than ancient proofs doesn't deny the insight of Gautama [a.k.a. Buddha] or any of the gifted writers and poets of the many paths. . . . We're just practical about the spiritual needs of a twenty-first-century globally interconnected world."

The theologian Brian Alston has a dissenting view. He recently published a pamphlet entitled *What Is Neurotheology?*, in which he said that the new field has an essential flaw. It attempts to unify two different perspectives on human beings within one discipline, which is not possible because—citing the nineteenth-century philosopher Friedrich Schleiermacher—"science is accessed through knowledge and religion through feeling."

But Schleiermacher didn't know that one day we'd be able to see feelings at the moment they occur in the brain.

Aldous Huxley, the author of *Brave New World* and *The Doors of Perception*, used the term "neurotheology" in 1962, in the last novel of his career. *Island* is about Pala, an imaginary Pacific island whose people know how to merge tradition and science very effectively. The ruler of the island has a book that reflects Mahayana Buddhist principles and is called *Notes on What's What*. In his foreword for *Island*, Huxley described this fictional realm as a place where scientific and technological advances would be used as if they'd been created for mankind's benefit, not as if mankind were to be adapted to and enslaved by them.

I haven't met any researchers yet who believe they're going to crack open the ultimate mysteries of the universe or fully replace traditional religious studies. But they do think that they will advance our understanding of what it means to be a human in this cosmos, and possibly even help people realize how the attributes Abraham Lincoln called "the better angels of our nature"—such as compassion, forgiveness, love, and peacefulness—might be strengthened in the fabrics of our lives.

Right now, neurotheologians are not searching for the answers to the ultimate questions, but rather for the right questions to ask, and the best experimental designs for asking them.

Of course, religion opens up a supply of fascinating questions that are literally infinite. Stories of spiritual moments, religious epiphanies, and mystical experiences are consistent across cultures, across eras, and across faiths. Why is that so? If there is a God, did he or she hardwire our brains to believe in him, her, them? What is the neural nature of faith and belief?

There's an old joke: "What's the difference between the mind and the brain? The mind is what the brain does for a living."

Scientists wonder if that joke reflects a basic truth: Is our mind something that emerges from our brain, a sort of self-programmed software? Is it fully contained in the nervous system? Or is it a sort of pipeline to extraterrestrial territories, something like a crystal set, a radio receiver that pulls in additional frequencies and taps sources of information beyond our immediate five senses?

Some neurotheologians probing these questions expect to eventually prove the existence of God. Others expect to give scientific legitimacy to atheism. The neurologist James Austin is a devoutly spiritual person, while Matthew Alper, author of *The God Part of the Brain*, is an atheist. Both work to decipher whether God exists in our heads—or anyplace else.

Of course, using our brains to study our brains means that we are kind of boxed in. The limits of our brain and its ability to receive and process information are going to define the limits of our understanding. But it's still an amazing and enlightening journey, this fMRI view of our attempts to understand and feel connected to whatever is infinite. It may be the fullest expansion that scientific minds can make.

Many people around the world have no specific religion, yet practice various forms of spiritual seeking. There are others who take no stock at all in supernatural beliefs. There are nearly 2 billion people worldwide who self-identify as Christians, and a billion plus who are Muslims, in addition to more than 750 million Hindus, more than 300 million Buddhists, and over 14 million of Jewish faith. Those are tall numbers in any field. They represent billions of compelling reasons to use neurotechnology to learn what underpins all that belief.

We may eventually learn how to keep our religious natures acting benevolently, the way they are intended to work. A great many believers, while in high and powerful office, have taken the precepts of their faiths so seriously that disagreements about *who* God is and *what* God wants us to do have turned into major causes of bloodshed. When former president George W. Bush, whose brain's areas of speech formation seem hardwired for mishaps, applied the word "crusade" to American military involvement in Iraq, he evoked powerful resentments lingering from Christian-versus-Muslim warfare of centuries ago.

The dynamic between religious belief and science has also caused wars within cultures. Stem cell research is a frequent battleground of our time. Scientists accuse some religious adherents of preventing vital progress in medicine, while some religious people accuse the scientific community of "trying to play God."

Tensions between science and religion go way back, of course. The eloquent and challenging 2006 book *Cosmos and Psyche*, by Richard Tarnas of the California Institute for Integrative Studies, recalls both the intellectual thrills and the religion-based political dangers faced by Galileo and his sixteenth- and seventeenth-century contemporaries. They were convinced by the writings of Polish scholar Nicolaus Copernicus that our Earth revolved around the Sun, a theory that learned Greeks, Muslims, and Indians had already put forward long before. Copernicus had been reluctant to publish, knowing he would be despised by the religious authorities of his time. Martin Luther reportedly answered Copernicus's heliocentric theory with, "This fool wishes to reverse the entire science of astronomy; but sacred Scripture tells us that Joshua commanded the Sun to stand still, and not the Earth."

It's completely accepted now, of course, that Copernicus had it right. But the heliocentric concept butted heads with accepted belief, and people who embraced it were persecuted. Authorities of the day held that movements of objects in the heaven were too complex for human minds to understand. Case closed, thank you very much.

The theory of evolution, circulated by Charles Darwin nearly one and a half centuries ago, is the bedrock of the science of biology. Huge numbers of Americans still think that it is contrary to religion-based truths. Just like Copernicus, Darwin waited several years before publishing his findings, knowing full well that he would take a hit.

In Santa Barbara, California, at a November 2007 installment of an ongoing lecture series known as Mind/Supermind, Tarnas shared the stage with the humorist John Cleese—of Monty Python and *Fawlty Towers* fame. Cleese advanced his own definition of religion: "a primitive form of crowd control." He had spoken alongside Tarnas on the previous night at the Esalen Institute in Big Sur. The tongue-in-cheek title of his talk: "Why There Is No Hope."

Joking was set aside when Cleese asked Tarnas to talk about how we might reconnect to something valuable that primal human cultures hold, the sense that everything in the universe holds sacred value. It's a concept that lost traction after the emergence of scientific thought. Cleese clearly hoped that that kind of awareness might counterbalance some of the forces now threatening the continuance of human existence.

As they spoke, I thought about how neurotheology may become a radiant source of hope, a way of reconnecting with the sacred with the aid of science. However, I also believe the path leading up to that point in history is going to be rock-filled. As neurotheology progresses and as machines made by human minds are increasingly used to understand human minds, new flash points for controversy and political rancor will find their way into public debate. Resolving these conflicts without spilling blood, ruining careers, or suppressing research that could eliminate a lot of human suffering, is a vital collaborative job. It isn't likely to be easy.

Mike McCullough, a University of Miami professor of both psychology and religious studies, finds that it's tough to get scientists and theologians to agree on even definitions for basic terms like "religion" and "spirituality." But he sees tremendous potential value in getting past the conflict stage. A collaboration, even a fairly amicable truce between science and religion, could generate peace and save lives in several different ways, on a global scale. For example: If neurotechnology can tap into religious or even godlike experiences, what other emotions and feelings might we learn to utilize and control?

McCullough spoke at an October 2007 Claremont Graduate University conference entitled "Forgiveness, Generosity, and Sacrifice," a remarkable meeting of neuroscientists and professors of theology.

McCullough focuses chiefly on forgiveness and revenge, gratitude and religion. "The red thread that connects my current research," he said, "is the evolution of humans' moral sentiments and moral institutions." He draws on anthropological studies to show that the pressures experienced by a culture determine what its people will hold as ideals. Tribes and nations that have been pushed around tend to think of revenge as a supreme value. Cultures that are fairly prosperous and secure usually want transgressors who threaten their stability to be heavily punished. This underlines why religion, which is a holding place for values and ideals, and an area where people hope to find constant truths, is also a product of culture. While we're looking for a God who made humankind in his image, we keep finding examples of the process working the other way around: Humankind making gods in its image.

Andrew Newberg, of the University of Pennsylvania, directs the Center for Spirituality and the Mind there, and co-wrote a 2006 book entitled *Why We Believe What We Believe: Uncovering Our Biological Need for Meaning, Spirituality, and Truth*. He has been using an imaging technique known as single photon emission computed tomography, or SPECT, to study the brain.

Tomography is a way of taking multiple pictures revealing sections of a brain or other subject (the Greek word *tomos* can refer to a section or a cutting), which get pieced together by a massive computer to create a 3-D image. Newberg, though just forty-one, has been taking SPECT images for more than a decade. He has been trying to detect and measure what goes on inside the heads of people when they are immersed in a religious experience. The subjects have included Buddhist meditators, Franciscan nuns, and Pentecostal Christians who are able to self-induce the rapturous state of glossolalia, also known as "speaking in tongues."

Whether by chance or predestination, a woman on Dr. Newberg's research team happens to be a born-again Christian who experiences glossolalia. She told a *New York Times* reporter in 2006 that she thinks of her ability to speak in tongues as a "gift."[1] She remains aware of the environment surrounding her when she is entranced, and feels in control of herself—though not of the event. "You're just flowing," she said. "You're in a realm of peace and it's a fantastic feeling."

Newberg's team was the first to apply brain imaging to the practice.

There are two different forms of glossolalia. One of them is rather subdued. The other results in passionate outpourings that do not conform to any planet Earth language structures, but that are often very rhythmic. Glossolalia has been traced historically to Pentecostal churches established about a century ago, but may have deeper roots that were kept alive by slaves brought from Africa. Despite the horrendous repressive conditions they lived under, slaves generally had a fairly free rein with activities inside the confines of their churches. Christianity provided them with a matrix for preserving musical, spiritual, and other aspects of their African cultures. For example, slave owners had forbidden drums after they'd realized drumming was effective for communicating across long distances, and could be used to coordinate uprisings. Stomping in unison on a wooden church floor was a way to keep rhythmic elements of African cultures alive.

Newberg's team studied five women, including the born-again Christian researcher. All of the subjects were physically healthy, and all were regular churchgoers. Images were taken of each woman while she sang a gospel song, and again later while she was speaking in tongues. SPECT images showed their frontal lobes—where the brain does much of its willful, "executive" work, processing information and maintaining control—were relatively quiet while the women spoke in tongues. Amazingly, and ironically, so were the brain areas usually involved in language. Regions that maintain self-consciousness remained fully online. The brain region that became less active is one that is important for motor control and emotional control. It could be that relaxation in this area allowed the women to literally "let loose," escaping the constraints of some of the social conditioning written into their brains about "proper" behavior, but in a way that is admired within their religious community.

This brings up an important point. It isn't actually mandatory or predestined for science and religion to be in conflict. They have fundamentally different ways of looking at things, but they can still find areas of relative agreement. Newberg's brain scan images didn't contradict what the women believed was going on, the idea that God was speaking through them. They simply documented some biological facts about what happened in their brains at the same time they had their religious experience—the increased or decreased blood flow in specific brain regions. So it wasn't an either/or situation, but rather two different and nonconflicting ways of describing the same experience.

Imaging results with meditators and nuns show that their type of spiritual activity creates very different patterns inside their brains, compared to the patterns of people experiencing glossolalia. Meditators and nuns showed a lot of frontal lobe activity, which is typically a result of mental concentration. In another part of the brain, where sense input gets combined to help us see ourselves as separate beings, and to coordinate our movements through our environment, activities quieted down considerably. Newberg thinks the slowing down in this region allows people who are meditating or praying to feel merged with their activity, to feel that they and the meditation or the prayer are joined into a larger whole.

Newberg's findings are consistent with those of Dr. Les Fehmi, a psychologist and biofeedback expert whose 2007 book, written with Jim Robbins, is *The Open-Focus Brain*. Fehmi's methods make no reference whatsoever to religion. They are aimed at synchronizing brain wave activity, inducing a systemwide production of alpha-wave activity. The method involves guided meditations that instruct a hearer to notice feelings in various parts of his or her body, as well as spaces within their body and spaces outside, ranging from nearby to infinitely distant. For example, a listener might first be guided to quietly notice her awareness of sensations within her thumbs, followed in a few moments by noticing awareness of the presence of her forefingers, followed by sensing awareness of the absence, or space, in between thumb and forefinger. With that opening series as a template, the listener is guided in feeling awareness in various other places, both inside and outside her body.

The foundation of Fehmi's system is his belief we are culturally trained to focus our attention too narrowly. The result, he believes, is our minds suffer in much the same way a muscle does when it stays clenched. Widening of the attention is intended to relax the mind while still keeping it engaged. Although Fehmi and his associates have not tested their methodology with fMRI, they do use headbands filled with EEG electrodes to measure wavelength synchronicity throughout the brain. The equipment in Fehmi's lab emits a faint beeping sound when a subject's brain is in an alpha-wave state, a state associated with alert restfulness and letting go of tension. Subjects can work to increase the time spent in the alpha-wave state by trying to reproduce the beeping noise.

A person who had training sessions with Fehmi told me that by his third half-hour session the machinery indicated an alpha-wave state almost the entire time. According to Fehmi, many Open Focus practitioners report transpersonal (out-of-body or outside-of-time) experiences. These are not so different from what the meditators and nuns in Newberg's studies reported, but with religious terminology optional. It would be fascinating to see what fMRI would reveal about which areas activate and which deactivate during Open Focus exercises.

The nuns brains in Newberg's tests had increased activity in language regions, which makes great sense. They were focused on the words of their prayer. The Buddhist meditators were focused on visual images, and consequently their brains got into higher gear in visual centers.

Research similar to Newberg's has been carried out by Dr. Mario Beauregard of the University of Montreal, with Carmelite nuns as subjects. He and his team did fMRI scans on a group of fifteen cloistered Carmelites ranging from twenty-three to sixty-four years of age. Rather than trying to induce a religious state—something the nuns didn't feel they could do at will—the researchers asked them to relive a previous mystical experience. Scans showed that a dozen different brain regions became involved.

Brain region activation and deactivation is the center of an incredible event that was not an experiment that a scientist made happen, but rather something dramatic that happened to a scientist. Jill Bolte Taylor describes in her 2008 book, *My Stroke of Insight*, how the left side of her brain suddenly shut down one morning in 1996 as a result of a stroke she had suffered.

She was then thirty-seven, doing life sciences research at Harvard. A blood vessel in her brain ruptured while she was getting ready to start her day. A pool of blood formed, causing terrible pain and severely inhibiting the functions of her brain's left hemisphere.

In very general terms, the left side of our brains is more concerned with the rational and analytical aspects of life, and the right side with the emotional, creative, and intuitive. Left-brain activities are dominant in our culture. Taylor's mystical experiences, which she can discuss in scientific terms, suggest that this dominance may block our potential for happiness.

Taylor describes how the initial pain was followed by a sense of freedom, as if her mind had been unplugged from her body. Her internal dialogue about life's concerns and busyness went away, leaving her with a sense of relief so deep that she now calls it nirvana. She was, essentially, living only in her right brain.

A dozen years later she went back to academia again, this time as a teacher in the Indiana University Medical School instead of as a Harvard researcher. She has very consciously chosen to keep her right brain active. She claims to have developed the ability to shift her thinking away from the left brain whenever necessary, whenever it seems ready to impinge on her contentment. Though she might have stayed in a blissful but nonfunctioning state, she worked her way back because she wanted to let other people know that it's possible to be more at peace.

Taylor spoke at a February 2008 conference known as TED, which stands for Technology, Entertainment, Design. According to the *New York Times*, a video of her speech that went on TED's Web site has been watched by some 2 million viewers, and continues drawing about twenty thousand hits daily.[2] *Time* named her one of the world's one hundred most influential people, and Oprah's Web site posted its own Taylor interview. "Nirvana exists right now," Taylor told the *Times* in an interview when her book was released. "There is no doubt that it is a beautiful state and that we can get there."

From the perspective of evolutionary biology, the brain has two main jobs: self-maintenance and self-transcendence. In other words, selfishness is necessary to some extent for survival, but it presents a tough quandary. It can also become self-limiting and self-defeating. Israel Zangwill, author of the 1908 play *The Melting Pot*, which provided America with one of its best self-identifications, expressed these countervailing brain jobs like this: "Selfishness is the only real atheism; aspiration, unselfishness, the only real religion."

Newberg quips that God won't go away, because our brains wouldn't allow him to. He believes that religion evolved as a tool for helping the brain balance its dual workload. It may be meant to keep us aware, through its rituals, observances, and occasional mystically interpreted experiences, of a bigger perspective: Selfishness helps you make your way through this world, but too much selfishness is a self-defeating mistake. Maybe that's why the satirist Ambrose Bierce defined "alone" as "being in bad company."

Evolutionary biologists focus on the reasons why organisms change over time, and one of their key interests is sociocultural change. They are fascinated with the why of religious belief, and have theorized what goes on in our brains might prove evolution sometimes rewards individuals who cooperate with others. They think that religion may be an outgrowth of that trait.

Newberg's 1999 book, *The Mystical Mind: Probing the Biology of Religious Experiences*, was co-written with the late Eugene D'Aquili, who was also affiliated with the University of Pennsylvania. In it they said that two classes of neuropsychological mechanisms are the foundation of development for religious experiences and behaviors. In talking about these mechanisms, Newberg and D'Aquili used the terms "causal operator" and "holistic operator" to define two different networks of brain regions that perform a specific task. The causal operator sees close up, and focuses on how one thing leads directly to another. The holistic operator is for distance vision, allowing us to pick out the patterns within diverse happenings. This notion of two operators helps explain how we constantly make choices between very selfish and highly unselfish motivations.

D'Aquili and Newberg sketched a continuum of possible states of consciousness, from a baseline state—only taking in what is immediately experienced—to what they called Absolute Unitary Being (AUB). In the state called AUB comes a complete loss of the sense of having a separate self in space and time, so that everything seems undifferentiated and infinite. People who have experienced AUB, even those who were both highly educated and materialistically oriented prior to feeling this mystical unity, come to feel that the selfless, timeless state is more "real" than baseline reality. However, D'Aquili and Newberg didn't take sides over which place on the consciousness continuum is best, saying that each state of mind is real in its own way, and each in its own way serves us in adapting to life.

All of this leaves a fundamental question unanswered: Does a certain kind of euphoria come from a nonmaterial source, from a creator? Newberg wryly observes that the most amazing scan of a person absorbed in feeling a religious experience would be one that showed no different brain activations at all. That would be extremely hard to explain in purely scientific terms.

Until an experiment like that is recorded, we're left to ponder the fact that intensely euphoric religious experiences arrive with certain neurological markers, biological tracings that can be seen in our brains.

A fascinating thing about brain activations and religious experiences: Very similar things happen during other kinds of euphoria. The comedian Franklyn Ajaye tells a beautifully rambling joke in which God keeps telling his angels to add thousands and thousands more nerve endings in the genital areas of his man and woman prototypes. Finally some angels ask, "Why so many?" and God answers, *"I want to hear them calling my name."*

The brain is actually a fairly energy-efficient mechanism. One of its strategies is to use the same areas for similar functions. Euphoria can result from several different kinds of stimuli, and can be experienced in a wide spectrum from mild to intense. Both amateur and professional athletes and coaches expend huge amounts of physical and mental effort to win games, series, and championships. Part of their payoff is in the high postvictory spirits, and lasting emotional satisfaction, which they often seem to value above money. Veteran free agents who are courted by several teams will often take millions less in order to join a group with a legitimate chance to win a title. Even when the elation of winning ends up costing millions in lost salary, coaches and athletes seldom regret their choice.

This doesn't solve the mystery of transcendent and mystical feelings, but it does open our awareness. We know religious feelings, and a search for life's meaning, are common to much of humanity, maybe even to every single one of us. If we keep finding similar brain activation patterns occurring for similar kinds of activities among the world's many different religious and spiritual practices, we may find it easier to accept cultural differences, allowing that all these different kinds of transcendent journeys are ultimately headed for a single destination. William James wrote in his 1902 book, *The Varieties of Religious Experience*, based on lectures he had given at the University of Glasgow, "The divine can mean no single quality, it must mean a group of qualities, by being champions of which in alternation, different men may all find worthy missions."

Polytheistic religions might simply be using a group of deities to represent a group of qualities. In pre-Christian Hawaiian belief, each separate deity had both male and female manifestations. According to the kahuna (the word can mean something like "specialist" as well as meaning "priest") Aupuni Iw'iula, the current bearer of a two-hundred-year-old family tradition, there is also a little-known and seldom-used ancient Hawaiian name for a single, all-inclusive God. Eventually, neuroscience is likely to prove that the world's different religions have a lot more in common, which might pave the way for greater toleration.

Because it causes such dramatic things to happen in the brain, the disease of epilepsy is a gateway for neurotheological studies. An estimated 50 million people worldwide have epilepsy, which can come and go during childhood but can also persist across a lifetime. Since early recorded history, people have made a connection between epilepsy and religious experiences, from the most pleasant to the most terrifying. Epilepsy was called "the sacred disease," and people believed that epileptics could progress through the disease to having shamanic powers.

In 1936, when an English girl named Ellen White was nine years old, she got chased by an older girl while walking home from school. When White looked over her shoulder to gauge whether her tormentor was getting close, she tripped and smacked her nose on a stone. Out went the lights for the next three weeks. When she came to, she believed the shape of her face had changed. She was never able to return to school, and her personality transformed. Eight years later she began experiencing visions. Sometimes her expression would suddenly change, and she would stare upward, apparently unaware of her surroundings, sometimes making repetitious movements or gestures that she wouldn't remember when the event was over. Scholars suggest these movements describe temporal lobe epilepsy, but White wrote some one hundred thousand pages about her beliefs and became one of the founders of the Seventh-Day Adventist Church.

Most Seventh-Day Adventists reject the epilepsy idea and believe instead White experienced a truly divine connection. Science sees epilepsy more prosaically, as the result of brain circuitry changing behavior due to electrical overload. However, science also notes a fascinating connection between epilepsy and religion. A study reported at a 2002 American Neurological Association meeting, produced by Dr. Thomas Hayton of the New York University Hospital and his associates, described giving ninety-one epilepsy patients a standard questionnaire on spirituality and religious beliefs. The subjects reported a higher than normal likelihood to experience religious feelings, like a sense of God's presence, inner peace, awe in the beauty of life and creation, and love. Their scores in measurements of strength of belief and values also ran higher.

In a 2003 study, two professors of neuroscience at the Norwegian University of Science and Technology, Doctors Asheim Hansen and Eylert Brodtkorb, examined eleven peculiar patients with epilepsy. The patients reported erotic sensations, hallucinations, religious and spiritual experiences, and other symptoms that were beyond their power to describe. Eight of them actively wanted to keep on experiencing seizures. Five of them had the ability to self-induce a seizure, and four deliberately failed to comply with treatments, keeping their pipeline to ecstatic experiences open and flowing.[3]

Just over a decade ago, studies probing the connection of epilepsy and spirituality generated headlines saying that researchers had found a "God module," a single brain area for religious experiences. Those news reports were overwrought. The research team, directed by V. S. Ramachandran, emphasized that it was releasing preliminary data, and follow-up would be needed to verify its findings. "There may be dedicated neural machinery in the temporal lobes concerned with religion," said the report. "This may have evolved to impose order and stability on society."

The researchers never actually used the phrase "God module." But the concept was exciting, and the media's "newsworthiness module" lit up around the speculation. Since then, tests have shown repeatedly that several areas are involved in religious experiences, and they fire up to different degrees and in varying combinations, depending on a lot of variables.

Ramachandran's group studied a group of epileptics who reported that religious raptures occurred along with their seizures. The team found that mystical episodes and intense devotion to spirituality correlated strongly with epileptic seizures in the temporal lobes, a group of neurons above your ears. The theory that emerged was that electrical overstimulation of this circuitry produced the experiences. In fact, the researchers found that by mildly zapping the temporal lobes of some subjects they could induce feelings of a supernatural "presence." The key, Ramachandran speculates, may be the limbic system, interior regions of the brain that govern emotion and emotional memory. Epileptic activity may spark religious feelings because of electrical impulses that strengthen the connection between the temporal lobe and these emotional centers.

The fact that there is not a single "God spot" is actually not a scientific disappointment. It reinforces just how important religion and spirituality are to our brains. Any function that is highly important—such as music or language—will involve many regions of the brain, which eventually have to combine their information in ways that help the person receiving the input to make sense of his experience. The brain uses this strategy so that damage to one part of the brain won't necessarily cripple the whole system or even stop a certain function. If there were just one "God spot," a single injury could disconnect us from something we humans have found irresistibly important since ancient times.

Echoing Ramachandran's discovery of what electrical pulses to the brain can do, some surgical treatments for epilepsy call for implanting electrodes in a deep brain region. In October 2006, the *New York Times* reported that mild electrical stimulation through such an electrode gave one woman the sensation that she was outside her body, hanging from the ceiling and looking down.[4] Again, that old transpersonal magic. Another patient felt as though someone stood behind her, wanting to interfere with body movement. Dr. Peter Brugger, from the Neuropsychology Unit of University Hospital, Zurich, speculates, "The research shows that the self can be detached from the body and can live a phantom existence on its own, as in an out-of-body experience, or it can be felt outside of personal space, as in a sense of a presence."

Just outside of Ontario, Canada, similar experiments were begun in the late 1970s by Dr. Michael Persinger, organizer of the Behavioral Neuroscience Program at Laurentian University. Done in the days before fMRI, these experiments were some of the earliest attempts to find common ground in chemistry, psychology, and biology.

Persinger's experiments directed low-strength electrical pulses to three very significant brain areas using what he called "a God helmet." It excited a wearer's brain with electromagnetic bursts that mimicked the brain waves an epilepsy patient experiences during religious visions. The God helmet, a vivid shade of yellow that suggests it may once have belonged to either the Oregon Ducks or the Green Bay Packers, repeatedly produced spiritual-seeming responses in subjects, including hallucinations and visions, out-of-body experiences, and sensed presences.

Interestingly, Persinger pretested his God helmet subjects with a psychological questionnaire intended to reveal the degree of sensitivity they had in their temporal lobes. When he tried the God helmet on the famous atheist Richard Dawkins—author of *The God Delusion*—all Dawkins got from the experience was a headache. Persinger then pointed out that Dawkins's score of temporal-lobe sensitivity placed him in very low percentile. This suggests that Persinger may have discovered a neurobiological indicator for atheism. Or, looking in the opposite direction, that a brain with highly sensitive temporal lobes would be easy pickings for religious teachers.

In the years to come, it will probably be commonplace for people who haven't yet devoted years of their life to meditation and prayer to achieve mystical states with the aid of a device. The point would be not the mystical experiences themselves, but the significant and possibly lasting aftereffects: release from depression, better immune functions, a more positive overall outlook on life.

In fact, studies have shown direct health benefits from religious and spiritual practices. William James, who battled depression in his younger years, once wrote, "Believe that life is worth living and your belief will help create the fact." James's thought syncs up with the fact that meditation practice can progressively build up resistance to stress-induced illness. Lower blood pressure, lower heart rates, decreased anxiety, and decreased depression have all been demonstrated. In this regard, scientists now believe something that Buddhists have believed for a long time. We all function with certain "set points" in our internal thermostats, so we all have our tendencies to bigger or smaller amounts of anxious feelings, depression, and so on. Practicing meditation may act like hiring a technician to access the control panel and slowly dial in better settings.

Alan Wallace, head of the Santa Barbara Institute for the Study of Consciousness, holds a doctorate in religious studies and is also an ordained Tibetan Buddhist monk. I once attended a lecture where someone asked him, "Have you become enlightened?"

"No," Wallace answered, "but I have changed in many ways, and have gradually gotten away from some tendencies that I'm very glad I could leave behind." Similarly, Dr. Elizabeth Garcia-Gray, psychiatrist and past president of the American College of International Physicians, compares meditating to "running a defragmentation program on your computer."

Some spiritual practices aren't so likely to meet wide approval. Mind-altering drugs and religion have a long history that continues today in rituals throughout the world. Drugs with these powers can bring about horrifying experiences as well as ecstatic ones. Traditional cultures still have bodies of knowledge to guide people through or around difficult experiences, toward the possible spiritual payoff. Many modern cultures don't.

Somewhere in that large problem there may be clues to a solution. It could be that drug abuse is a sadly mistaken form of reaching for the transcendent. Neurotheology would like to know more about mind-altering drugs, and how they can cause religious experiences. The ultimate prize may be tools that eliminate the peril but still deliver the payoff. If so, the enormous social problem of addiction might gradually become a relatively small one.

Professor Roland Griffiths of Johns Hopkins University reported a study in 2006 in which sixty volunteers who were interested in religion and/or spirituality were given either psilocybin—the psychoactive ingredient found in some mushrooms—or Ritalin—a drug used to focus brain activity in people with attention deficit/hyperactivity disorder. The volunteers went to two sessions, eight hours each time, with a sixty-day interval in between. They weren't told which drug they were receiving, but were asked to describe thoroughly any immediate or long-lasting effects. Ten of those who were given psilocybin said their experience was the most significant spiritual experience of their lives. Twenty of them placed it among their five most meaningful life experiences. More than half called their psilocybin episode "a full mystical experience." What they described matched almost perfectly descriptions given by people who attained their mystical experiences without drugs. The subjects who took Ritalin did not report any mystical experiences.

Griffiths is careful, as anyone ought to be with one foot in the minefield of religion and another in the IED Alley of psychoactive drugs. Knowing that the study could fuel controversy about the existence of God, he and his associates wrote in the journal *Psychopharmacology*, "This work can't and won't go there." But they did assert that "under defined conditions, with careful preparation, people can safely and fairly reliably occasion what's called a primary mystical experience that may lead to positive changes in a person. It is an early step in what we hope will be a large body of scientific work that will ultimately help people."

Ramachandran sees promise in mirror neurons, which cause the mirror-touch synesthesia described in the chapter just before this one. We all have mirror neurons, and Ramachandran says that they are "the key to understanding human perception and action.

"It almost seems like science fiction," he shares. "When I touch you, the neuron fires. The same neuron fires when I touch somebody else. But if I touch somebody else, you will not *feel* it. And that made me wonder why.

"I believe that this is the answer: You've got your real skin receptors telling several other touch neurons in your brain, ones which are not mirror neurons,

'Hey, I'm not being touched.' Then that message vetoes the output of the mirror neurons.

"Now, some of the output gets through, of course. That's why you can empathize and say, 'Oh. He's being touched the same way I'm being touched.' "

Ramachandran got the idea to study amputees, patients who had previously lost arms or legs and therefore no longer had skin receptors in certain parts of their bodies. "Believe it or not, and I know this sounds like *The X-Files*," he says, "if you go and touch somebody else, this guy feels it on his missing limb. You rub the other person; he feels his phantom limb being rubbed. You poke the other person; he feels his phantom limb being poked. So, here's a neuron which literally, in this phantom-limb business, dissolves the barrier between other human beings and you. And these neurons exist for emotions and for physical pain. I call them 'Gandhi neurons.' And I can't think of anything more important for society, especially in this day and age, with all the terrorism and war happening."

Understanding these Gandhi neurons and mirror neurons could turn out to be critical to understanding social empathy, which could lead to dissolving barriers between people. That's a goal of most of the world's religions, if not all. Many religious traditions assert we are all really one.

Ramachandran sees possible applications for these Gandhi neurons in treatments for mental illness. One of the cruelest aspects of depression is a sense of aloneness, an inability for the depressed to relate to or think about much of anything except their own troubles. Anything to enhance feelings of connectedness could be powerful therapy.

"We need to do research with drugs like Ecstasy, which I've long thought can enhance empathy. It has other side effects and clearly the molecule needs to be modified. But it could be an empathogen, a substance to help us become empathetic. It probably works by stimulating empathy mirror neurons. So it's possible that pharmacological or other interventions based on this research could help some people enjoy better mental health. Or maybe with more direct interventions you might be able to enhance that activity, and thereby make people have more empathy."

As my conversation with Ramachandran draws to a close, I suddenly recall a fascinating thing that happens every weekend in San Francisco, about three miles from my neighborhood. A church meets for services that turn into free-flowing jazz jam sessions. Members of the church are inspired religiously by the late saxophonist John Coltrane, and they see him as a saint.

Coltrane's best-known work is *A Love Supreme*, recorded with his quartet in 1964 in a single session. Jazz lovers almost unanimously put it on their short lists of the greatest albums in jazz. It's a four-part suite, incredibly rich harmonically, with tracks named "Acknowledgement," "Resolution," "Pursuance," and "Psalm." People in the church call it a thesis, not an album, and say on their Web site, "We thank God for the anointed universal sound that leaped down from the throne of heaven out of the very mind of God and incarnated in one Sri Rama Ohnedaruth, the mighty mystic known as Saint John Will-I-Am."

What got me thinking was the fact worshipers at this church find their pathway to religion through music. As soon as I tell Ramachandran about this, he feels the story click with what he has been researching over the years. "We don't understand it, but I think music and visual art can take you into a transcendental plane. And there you'll start merging with religious experience."

When we peer deep into our emerging neurosociety, we can see that there are many ways neuroscience will impact the religious and spiritual lives of billions of individuals across the planet. While the advent of drugs and devices to safely stimulate out-of-body and spiritual experiences will capture the attention of certain spiritual seekers, others will leverage neurofeedback technologies like real-time fMRI to accelerate their natural capacity to achieve elevated experiences that transcend self-seeking boundaries. This technology can provide visual guidance to attain an indescribable state of mind, like a map shows the best way to a previously unknown place.

At a global level, the repeated neurotheological exploration of spiritual, religious, mystical experiences across the world's religions will reveal we all share common moral instincts and intuitions, such as fairness and empathy. These unifying threads will be uncovered and promoted widely as a glimmer of hope for humanity as we wade through the increasingly contentious religious future. Looking forward, neurotechnology will significantly alter how people view their faith, spiritual existence, and the culturally created world around them, directly challenging specific aspects of traditions of many world religions. The rise of the neurospiritual tradition will naturally further escalate the rigidity of particular reactionary groups who hold tight to their preneuroscientific beliefs.

But just as Copernicus's heliocentric notion of the universe is now bedrock truth, the Neuro Revolution will bring about new ideas of human spirituality that will forever reshape our understanding of humanity's role and place in the universe. A quiet transformation has begun, albeit one that may take centuries to play out fully.

EIGHT

FIGHTING NEUROWARFARE

The arms race is based on an optimistic view of technology and a pessimistic view of man. It assumes there is no limit to the ingenuity of science and no limit to the deviltry of man.

—I. F. Stone

It's God's job to judge the terrorists. It's our mission to arrange the meeting.

—U.S. Marine's bumper sticker

In just a few paragraphs, the incense and flowers and music of hippie-era California will fill the air. But before we trace the neuroscience- and war-preparedness-driven events that led to a countercultural explosion, these few quick historical vignettes will demonstrate why we need to think about the inevitable future interplay of warfare and neuroscience.

A lilting song that nearly every kid learns, "London Bridge Is Falling Down," actually commemorates a military raid that destroyed a bridge over the Thames. The Vikings were able to terrorize England, Ireland, Scotland, and many other places because of their superior technology in boatbuilding. Viking ships were lightweight, and built so they flexed along the keel line. This lightness and flexibility gave them speed over rough waters, and also enabled a tricky tactic. The raiders would tarp their boats, covering themselves under hides and blankets, then purposely take on water. When their craft became barely visible above the surface, they would row ashore, drag their boat onto land, make a lightning strike, and scoot back out to sea before local defenders realized what was happening.

They also frequently left behind genetic material, to state it delicately, which still shows up in the lighter-complected Scots, Irish, and English of today.

On land, other nations made technological breakthroughs in warfare via the training of horses and of infantry soldiers. Later, in bigger boats than the Vikings had, outfitted with cannon, nations with the strongest naval forces dominated. Still later came the war-waging advantages of rail, motorized ground and air vehicles, radio transmission, satellite transmission, more powerful ordnance, and most recently, the laser-guided ability to send a bomb practically into your enemy's sock drawer.

Now the frontier is the human central nervous system itself, as connected to, enhanced by, or defeated by emergent neurotechnology. We've already been exploring this frontier for a few decades, and we've already seen a gigantic cultural spin take place, significantly charged by materials that came out of neuroscience research.

In the spring of 1965, under abundant rays of sun in Palo Alto, Stanford students are dancing to live rock and roll by the Warlocks on the rooftop of Tressider Student Union.

Warlocks fans are knocked out by the lead guitarist's lengthy exploration of a steady-rocking chord progression. He spins out a multimodal solo of the kind novelist Ken Kesey will later describe as the musical version of a snake slithering through a woodpile. These hypnotic musical snake dances will soon earn this guitarist the nickname Captain Trips.

The Warlocks are electric in every sense of the word, though until pretty recently the core members—billed as Mother McCree's Uptown Jug Champions—played acoustic folk-style tunes in a local bookshop. Within the year—reportedly while smoking DMT, a hallucinogen prized for its ability to bring on mystical and spiritual experiences—they will reincarnate as the Grateful Dead, and become the "house band" for Kesey's infamous Electric Kool-Aid Acid Test parties.

Kesey will bankroll his LSD bashes with royalties from his 1962 novel *One Flew Over the Cuckoo's Nest*, a book he wrote following extensive experiences in psychopharmacology—including DMT, psilocybin, mescaline, LSD, and cocaine—which began when he was a human guinea pig for Project MK-ULTRA, a CIA-financed study into mind-bending chemistry, conducted nearby at the Menlo Park Veterans Hospital.

Later, in 1975 and via the Rockefeller Commission Report, Americans would learn that their government had frequently dosed members of the general public, as well as prostitutes, mental patients, and various people in the military, doctors, and CIA employees. Most of the time the subjects weren't informed.

Radical experimentation in drugs, lifestyles, clothing, music, and protest-centered politics had already begun before the casual dance at Stanford, but it was then about to begin spreading quite rapidly through California youth culture and beyond, like "ripples in still water," to cite a Dead lyric. According to later estimates, the soon-to-be-famous chemist Owsley Stanley would eventually create some 5 million hits of LSD, at 100 mgs apiece.

In early 1970, when Ronald Reagan employs his twinkling eyes and avuncular charm in the role of governor of California, he will tell some agriculturalists gathered at Yosemite's historic Ahwahnee Hotel that if a bloodbath is necessary to quiet protests by campus radicals, he is in favor of its happening right away. "Appeasement," he asserts, "is not the answer." Later, though, he will appease people who didn't so much want America's collegians eating hot lead, claiming he'd meant "bloodbath" as "a figure of speech."

Everyone knows that the meet-up of psychedelic drugs, politics, and popular culture spawned unusual tales. What is less known is that these stories of side effects and unintended consequences could be labeled "Your Tax Dollars at Work."

Psychoactive-drug research was launched by America's quest for technological edges in warfare. Political and military leaders of the day apparently didn't imagine the outcome, but the psychedelic movement and its apostles—Timothy Leary, the Beatles, the Rolling Stones, the Jefferson Airplane, Steppenwolf, Jimi Hendrix, Charles Manson, et al.—came into their full flower-power bloom largely because those fascinating psychedelic drugs in government research programs became enormous popular hits, in at least two senses of the term. Had the American government arranged to derive a percentage from every song, book, or movie inspired as a consequence of their research into psychoactive drugs, our national debt might now be a distant memory. There was so much casual drug use in society, some psychologists believed that any young people who hadn't yet experimented with psychotropic substances probably had constricted personalities. Or, as Dylan sang, "*Everybody* must get stoned."

Ironically, when the government quickly and reactively outlawed LSD on October 6, 1966, research projects that might have mattered a great deal were permanently shut down. The potential for liberation of consciousness—an idea that attracted people so powerfully that they risked their brains on homemade drugs from freelance chemists—might have been realized by now, as researchers of the time hoped, in medicines for the millions of people who suffer from depression or other mental illnesses.

After all, the pipeline for the transfer of military knowledge for civilian benefit is a venerable one. World War I brought about the National Research Council, established in 1916 as a part of the National Academy of Sciences, which was chartered by Abraham Lincoln in 1863 to be our government's main source of science, technology, and health policy advice.

The loss of the healing potential in now-banned psychoactive drugs not only vexes, it also stirs concern about what might go wrong in present days, as neuroscience and statecraft bounce in the same bed, surrounded by far more dangerous toys. The Defense Advanced Research Projects Agency Web site features a link near the top of its home page that inquires, "Are you a scientist or engineer with a radical idea (or ideas) that you believe could provide disruptive change for the United States military? Find out more. . . ."

And, in fact, the preponderance of the research now creating our emergent neurosociety is underwritten by America's defense spending. The neuroeconomist Read Montague of Baylor University won't take money from DARPA, but he represents a tiny minority. According to the Association of American Universities, nearly 350 colleges and universities held Pentagon research contracts in 2002, representing 60 percent of basic research funding. The leader in 2003 funding was MIT, which drew a half billion dollars. As a member of the leadership board at the McGovern Institute for Brain Research at MIT, a new $350 million brain-research facility, and from my visits to many other universities and private labs around the country, I know there are many of brilliant scientists who have both the ideas and the willpower to accelerate radically our understanding of the brain, if given sufficient financial support.

Meanwhile, a staggering amount of innovation has already arrived as a direct result of DARPA employment and research grants. That's exactly what was intended when the agency was formed in 1958 (until 1972 it had the shorter acronym of ARPA). It was a response to the shock wave of fear Americans felt when Russia surprised us by grabbing the first big headlines of the space age with the launch of its Sputnik satellite. There was an urgent national sense that we had to buckle down, and speed up America's science learning in particular. Some communities built new schools that were devoid of windows, each classroom just three cinder-block walls facing a blackboard and a podium, as if daydreaming could be eliminated so America would get back on top.

When ARPA began missile-tracking research, it asked scientists to figure out how computers in different locations could talk to one another. The resulting linkage was at first called the ARPANET. Today we just call it the Internet. We click on to it with the computer mouse, another outcome of DARPA-funded study. More key pieces of technology resultant from the massive (estimated around $3.2 billion) budget of the agency include the M16 rifle, stealth jets, wearable computers, long-range drone aircraft, ground radar, the Saturn rocket, and night-vision scopes.

DARPA tries not to be overly secretive, perhaps taking risks about what an antagonistic country or stateless enemy might do with research results it can access quite openly, because DARPA is convinced that an agency culture that promotes sharing of discoveries will continuously spark the creation of new knowledge. Add to that DARPA's belief it has the momentum and personnel to stay miles ahead of the rest of the world.

What you will read in this chapter may excite you about possible future benefits from work commissioned by DARPA and similar agencies. But it may also evoke nightmares about possible side effects and future unintended consequences. Humankind's long-standing, sanctioned-from-the-top-down custom of mass killing and mass retaliation is being revolutionized by neuroscience, and it will gain otherworldly weapons and tactics.

Neurowarfare is already a fact. Its expansion is as inevitable as the massive changes coming in all the other segments of society. Therefore, this chapter may also give you strong feelings about who you would want to hold the commanding lead in acquiring and controlling this world-shifting neurotechnology.

Just as nuclear fission has already done, neurowar is going to create a perpetual state of tension between promise and peril. Worry, debate, and conjecture will tumble through our minds over what the ultimate effects will be.

For example, put aside all issues of ethics, practicality, and good judgment, and suppose that we could have pacified Osama and the mullahs, or Saddam and the Republican Guard, bloodlessly, by, perhaps, some kind of psychoactive invasion. What if we could have slipped them something that turned all their aggression into a desire to spend hours digging heavy metal through cranked-up iPods, or if we could have turned them into a bunch of very mellow dudes—or whatever else was on the CIA's wish list when it framed Project MK-ULTRA? Or suppose that limitless oxytocin, with a discreet but effective delivery mechanism, could inspire Sunnis to trust Shiites, could induce the world's willing-to-kill-for- and/or willing-to-die-for-the-cause fundamentalists to respect their counterparts all around the world? What if some product of neuroscientific research could convince all nations—including those labeled rogue states, the Great Satan, or members of the Axis of Evil—to eliminate entirely their weapons of mass destruction?

Imagine that degree of promise rooted in the same body of research that could also manifest weapons so gruesome we can barely conceive of them. When I began this book, I drafted a sort of overview sentence that went, "It may seem like we are about to fall into a chasm, but thanks to the curiosity and drive of our ancestors and billions of us working together today we may soon be able to build a bridge wide enough for all of us to survive." Of all the facets of the neurosociety that are now evolving, it's neurowarfare that most makes me hope my optimism is right on.

Governments with untold billions at their disposal are steadily increasing their investment in neurotechnology. We're on the cusp of eerie and disturbing developments that seem like they've been taken straight out of *The Manchurian Candidate*. Sophisticated neuroweapons for coercive truth detection and the erasure of memories are already on the horizon. Recently revealed Soviet research, conducted under the name Project Flute, included plans to deploy a neurotoxic agent that would remain dormant until activated by stress or great emotion, at which time it would damage the nervous system, alter moods, trigger psychological changes, and even kill. Project Bonfire, also developed in the cold war Soviet Union, manipulated peptides and hormones that regulate our nervous systems.

It's likely Pentagon officials have known about Flute and Bonfire for quite a long time, and have been working intensely to create equivalent measures and countermeasures. Will such developments provide technologically advanced societies with greater national security? Will they instead "level the playing field," allowing the tinhorn megalomaniacs on various continents to back their posturing with massively inhumane neuroassaults? The second battle of Ypres, fought over a five-week span in the spring of 1915, goes down in history as the first massive use of poison gas. It ended in a stalemate, but while the German forces lost some thirty-five thousand troops, Allied forces lost around sixty thousand. That's an impressive differential, enough to convince many people to overlook the international laws against chemical and germ warfare.

Warriors armed with advanced tools from brain research will eventually provide the answers to how neurowarfare will unfold, because, in fact, both the "good guys" and the "bad guys" are already eyeing the possibilities.

Russia's supersecret OSNAZ forces, a far more clandestine version of America's Green Berets, gave the world a preview late in October 2002, when they used neurowarfare to answer the challenge of some 42 Chechen militants who held a reported 850 men, women, and children hostage in a theater in the Dubrovka area of Moscow. The audience was watching a performance of *Nord-Ost*, a musical based on Russian history. During act 2 the militants came onstage and said that they were personally packed with explosives, and had stashed more bombs throughout the building. Everyone in the theater would be killed, the militants claimed, unless Vladimir Putin agreed immediately, unconditionally, to pull all Russian forces out of Chechnya. They supplied the media with a videotape saying, "We have decided to die here in Moscow. And we will take with us the lives of hundreds of sinners." Performers who escaped through a window backstage said that about half of the militants were women in burkas.

Early on the morning of October 26, two and a half days into a standoff, OSNAZ fighters pumped a gas through the theater's air-conditioning system. To this date, Russian authorities haven't said what they used. The consensus is that it was a relatively new form of a synthesized opiate called fentanyl, which hits the human nervous system with eighty times the potency of morphine. Whatever they deployed, it sedated many of the militants and hostages into deep sleep but killed many others. Although the terrorists were aware they were being gassed, and had several minutes in which to react, the threatened destruction of the theater didn't happen. Instead, OSNAZ fighters engaged in gunfire with militants who had apparently brought gas masks. The firefight, at first sporadic, turned intense when the soldiers blew open the main doors of the entry hall on their way to taking the building.

Reports from different sources are conflicting, and the Russian government suspended its investigation of the *Nord-Ost* siege in July 2007 with little to announce, but it's widely reported 33 militants and 129 hostages died. Of the latter, only 1 died from a bullet. The remainder were killed by what doctors diagnosed as respiratory depression. Many victims might have been saved. An antidote for fentanyl poisoning exists. But authorities would not identify the poison, so doctors on the scene could not administer an antidote.

Two days later, Russian soldiers killed 30 Chechen rebels outside Grozny. Four days after that incident, the lower house of Russia's legislative body, the Duma, approved stronger restrictions on press coverage of stories involving terrorists. In the following year, Chechens living in Moscow were harassed by police even more than they had already been, according to reports of Human Rights Watch.

Russia is a signatory to the Chemical Weapons Convention, which was introduced in 1973 and has to date been signed and ratified by 183 of the world's 195 nations. It is a disarmament treaty for chemical weapons, and it says any substance used for domestic riot control must have effects that disappear soon after exposure. Fentanyl, or whatever else the Russians may have used, clearly exceeds that treaty.

The *Nord-Ost* reaction highlights a fact that's even more important than a single treaty violation: Warfare tends to suppress rule-following tendencies. When the whip comes down, we have to expect dirty and ill-advised use of neurotechnology, whether it's in the hands of rebels or conservatives, outlaws or government officials. General George Patton famously declared the point of war is not to die for your country, but to make the other bastard die for his. The Chechen rebels probably would not have hesitated to use neurotoxins on the OSNAZ forces, given any opportunity.

Richard Nixon oversaw the dismantling of America's biowarfare research in 1969. It had been running since the 1950s, or the 1940s (depending on whose interpretation of still-classified information you consult). But according to stories pieced together from Russian defectors, our biowarfare team was playing touch football, comparatively, while the Russians were headed for the Super Bowl. They had four major anthrax facilities, located in Kurgan, Penza, Sverdlovsk, and Stepnogorsk. After Dr. Kanatjan Alibekov, a 1992 defector, was debriefed by Bill Patrick, a seasoned veteran of American biowarfare research, Patrick reportedly put his head down on the table where they sat and moaned, "Oh, my God. Oh, my God."

Alibekov now goes by Ken Alibek and uses his training to promote human immunity instead of destroying it. He was previously in charge of the Stepnogorsk facilities, which was approaching an annual production figure of one thousand tons of weaponized anthrax. Not only was Russia's anthrax production three hundred times greater than America's; its numerous scientists were also exploring breakthroughs that would make anthrax, botulinum toxin, and other "traditional" forms of biological warfare seem like relics of happier days gone by.

Sergei Popov, another ex-Soviet scientist who defected in 1992 and now works in the United States on health-promoting research, successfully modified a *Legionella* bacterium so it could cause severe nervous system diseases that mimicked multiple sclerosis. Once given conventional treatments, the illness would morph several days later into strange new symptoms. Popov worked in a Siberian facility with thousands of other scientists, several hundred of them with doctorates. He and his cohorts were constantly told that Russia was far behind the United States, and that they had to catch up aggressively. With multiple levels of secrecy in effect, most people were told only the "closed legend," a plausible story that concealed the true nature of each project.

The project named Bonfire toyed with the structures of bacteria to produce strains antibiotics couldn't cure. Another one, the Hunter Program, worked to create two-in-one hybrid viruses. The intent was to create self-timed biowarfare cluster bombs, such as bacteria that had viruses inside their cell walls. The bacteria would produce a heinous disease; then killing the bacteria would gradually release a viral infection. Popov's knowledge of Hunter is limited, but he knew they were working to put smallpox and ebola viruses inside the bacteria that cause bubonic plague.

In an article published a decade ago by the U.S. Army War College, military analyst Timothy Thomas used the title "The Mind Has No Firewall." It laid down this challenge: "This article examines energy-based weapons, psychotropic weapons, and other developments designed to alter the ability of the human body to process stimuli. One consequence of this assessment is that the way we commonly use the term 'information warfare' falls short when the individual soldier, not his equipment, becomes the target of attack."

Thomas pointed out that military strategists, at least up to the time of his report in 1998, had essentially looked only at simple deception as a means of toying with an enemy's capacity for rational thought. (This, of course, was before revelations coming out of the war in Iraq.) But, he says, the human mind and body must be seen as an information and data processor, and also protected somehow with a firewall. "The body is capable not only of being deceived, manipulated, or misinformed but also shut down or destroyed—just as any other data-processing system," Thomas commented. And, as neuroscientists have reliably proven, incoming data from the environment, such as electromagnetic and acoustic energy waves, like data originating within the mind and body through the body's own electrochemical responses, are subject to manipulation and change just like the data on an actual computer. Strobe lights, for example, can be used to induce epileptic seizures. In a strange incident a few years ago in Japan, many kids in various locales who were watching the same cartoon show either experienced seizures or otherwise became sick.

According to Thomas, a Russian writer named N. Anisimov, working for the horrifically named Moscow Anti-Psychotronic Center, came up with the term "psychoterrorism" to describe weapons that were being developed in the former Soviet Union. Anisimov defined psychotronic weapons as those that can remove, edit, and replace memories in a human brain. A former major in the Russian army reported in a February 1997 military journal that many weapons fitting the psychotronic definition were being developed throughout the world. Some were then already in prototype stages.

Thomas's article continues to provoke response, and is widely quoted by authors on Web sites that would seem, let's say, "excessively vigilant" to most of us, but research had already been carried out concerning possible USSR-versus-U.S. psychotronic warfare even before "The Mind Has No Firewall" was published. Disturbingly titled Project Pandora, the research was run by the psychology division within the psychiatry research section of Walter Reed Army Institute of Research. Pandora was initiated after we learned that the Soviet government had, from 1953 to 1976, beamed microwave radiation at the United States' Moscow embassy.

As I've pursued neuroscience topics over the past several years, the conferences where I get invited to speak have become more fascinating. In August 2007, on behalf of the Defense Intelligence Agency, I briefed a special committee on the current and future state of neurotechnology. The committee, convened by the National Academy of Sciences, has a name that you may be able to memorize someday, when you have two or three spare hours: the Committee on Military and Intelligence Methodology for Emergent Neurophysiological and Cognitive/Neural Science Research in the Next Two Decades. Or, in handier form, the CMIMEN&C/NSRNTD. Members of that committee want to identify trends in brain research that could help the U.S. intelligence community anticipate how far neuroscience may progress internationally by the year 2027.

About a dozen speakers came in for this intense two-day session. While I was waiting in the greenroom to make my appearance, a brief interchange alerted me I was stepping into an especially high-caliber meeting. A gentleman I subsequently learned was Master Chief Glenn Mercer, a twenty-year Navy SEAL veteran, heard me remark—for the sake of casual conversation—that the stainless-steel cart with two carafes of steaming brew rolling past us into the meeting "must have eighteen gallons of coffee." "Five gallons," he corrected, quickly. No smile, no continued conversation. Just a quick dose of reality. That's what we're here for. If you aren't 100 percent certain, mister, don't say anything.

Glenn was there to brief the committee on the neurotechnology needs for U.S. Special Operations Forces, or SOF. With over ten years of experience in forward operations, he now plays a leadership role in SOF's Human Performance Management Program. SOF includes the elite fighters from the four military components, Army, Navy, Air Force, and Marines. Glenn calls his comrades "the NASCAR of people, finely tuned engines."

He brought an incredible sense of urgency and realism to the discussions, focused on the needs of our country's "warrior-athletes" over the next five years. He didn't spin his wheels in any of the ethical or linguistic debates ping-ponging between committee members about whether the technology involved ought to be called human performance enhancement, human performance optimization, human performance enablement, human performance modification, or human performance "insert your own favorite noun here." He cared only about the actual impact of the technology on tactical capacity, the ability to fight effectively and efficiently. He described the troops' perspective: They go out in small teams, in the dark, to perform forty-eight-hour missions that sometimes include eight-hour excursions underwater. Their missions require packing forty-five pounds of gear and analyzing everything within five meters of themselves with life-or-death focus. They are a fraternal order, intensely loyal to one another, extremely motivated by patriotism, and able to smell threats from a mile away.

Many high-level athletes think of themselves, metaphorically, as warriors. But this analogy can also be turned around. Every warrior is an athlete, a package of learned and innate skills inside a human shell. Warrior-athletes need optimum conditioning and endurance in order to win—especially at the hyperdemanding performance levels of SEALs, Green Berets, and other elite fighting groups.

Our government invests a tremendous amount of energy, time, and money to bring these warrior-athletes up to speed. By the very nature of their job, every one of them will experience traumatic events. If any way can be found to preselect candidates who are better suited to such extreme conditions, or to increase the abilities of the people who will be going out on hellaciously tough missions, Glenn and his colleagues want to know about it. An area he spoke passionately about was identifying selective biomarkers of performance for cognitive fitness and stress resistance to help select and assess SOF candidates.

One biomarker test that has recently shown promise is for neuropeptide Y. Higher levels of neuropeptide Y have been shown to be associated with hardiness and resourcefulness, particularly on an emotional and psychological level. Such warriors would be more resilient when under stress, and more resistant to post-traumatic stress disorder, today's term for the anxious condition that used to be called "shell shock." According to a recent study by *Psychiatric News*, over 16 percent of Iraq and Afghanistan war veterans still suffer from PTSD one year after returning home. There are probably big numbers of unreported cases, because of the fear that seeking therapy could torpedo chances of future promotion.

The Department of Defense is also searching for biomarkers that will reveal naturally quick reaction time, high visual acuity, and the most comprehensive information retention possible.

Glenn described for the committee many other studies that have been performed, including the use of a vast array of nutriceuticals—natural nutrients that could improve the bodies and minds of warrior-athletes. Dehydroepiandrosterone (DHEA), the natural hormone that is the precursor of both androgens and estrogens, our male and female hormones, shows great promise for increasing physical and emotional resiliency, and there are several products on the shelves of health-food and nutritional-supplement stores that claim to boost sex hormone levels. Many promote the fact that they're made from wild yam extract, but it isn't yet certain human bodies can make use of hormones from wild yams. More vitally, there's strong concern that many hormone-sensitive cancers, such as ovarian, prostate, and breast cancers, could be stimulated by higher-than-normal levels of androgens and estrogens. Glenn says that the Department of Defense is now conducting the first long-term study of DHEA, and hopes to have soon a set of answers driven by the need for advantages in actual war, not just winning the battle of the commercial shelf.

Other studies in progress include research into increased alertness, pitting caffeine against dextroamphetamine, which goes by the trade name Dexedrine, and a more recently developed drug, called modafinil, which is marketed as Provigil for narcolepsy and excessive daytime sleepiness.

If you've ever put away too many espressos in too short a time, you know that lots of caffeine makes you jittery. But caffeine also yields a noticeable increase in mental alertness and short-term memory. The same is true of Dexedrine. Modafinil was first approved for use in Canada, in 1999, and has been used by America's troops in Iraq. It seems, short-term, to be free of adverse side effects. But withholding sleep, that sweet process that does so much to enhance mood and alertness, can give one's inner demons a free bump from coach to first class. Bodies that have done without sleep will revolt and throw their owners' minds into loops and spins. Whether this effect is completely stopped by modafinil is something only long-term review can establish.

This point was underlined shortly after Glenn's talk by Dr. Anjan Chatterjee of the Center for Cognitive Neuroscience at the University of Pennsylvania, who calls the use of performance-enhancing substances "cosmetic neurology." It may be a very apt term. Cosmetic surgery was pushed forward in the wake of World War I to help combatants who had been disfigured on the battlefield. As anyone who has watched the cable TV show *Nip/Tuck* is well aware, cosmetic surgery's main use today is in the reproductive wars, helping people to signal more youth than they actually possess. "No one has conducted thorough studies about how brain-boosting drugs would affect healthy people after years of use," Chatterjee told the committee. But a number of one-person studies are already under way.

Paul Phillips, a computer programmer who made the career switch to professional poker player, was elated by the official diagnosis of ADHD he received in 2003. He quickly filled a prescription for Adderall, a psychostimulant that increases levels of dopamine, the brain's chief feel-good chemical. Adderall is a mixture of amphetamines, so it has the potential for creating addiction. The drug made Phillips feel like an information sponge, absorbing the tactical subtleties of each opponent, crunching the input at real-time speeds to counteract their moves, and yet it also made him feel calmer, more patient, wiser in his decisions about whether to hold 'em or fold 'em. A year later he added Provigil to his chemical arsenal. In December 2007 he told Karen Kaplan and Denise Gellene of the *Los Angeles Times*, "There isn't any question about it. They made me a much better player."[1] He added that the drugs have helped him win more than $2.3 million at the tables. That amounts to a terrific study grant, and at the same time a dangerous linkage—cash flow and chemical highs.

One week after the *Times* story appeared, the *Journal of Neuroscience* printed a report from the Wake Forest University School of Medicine. Researchers there deprived monkeys of sleep, then gave them a naturally occurring brain peptide, orexin-A. Only a small number of neurons produce orexin-A, but it affects several brain regions. Its job in the brain is to regulate sleep. When you don't get enough sleep, your brain tries to generate more of this peptide. Eventually, it can't keep enough orexin-A flowing to keep you awake, and you nod off.

The monkeys were given a battery of tasks, for which they'd received prior training. Then they were blasted with videos and music, tempted with snacks, urged on by technicians, and kept awake for thirty to thirty-six hours straight. Then the same tasks were given once more. Monkeys who hadn't been juiced with medication did significantly worse, just as a sleep-deprived human would. Monkeys given orexin-A, either by a hypodermic injection or a nasal spray, did their tasks just as well as they had the first time around.

Samuel A. Deadwyler, a Wake Forest professor of physiology and pharmacology, drew the obvious conclusion: "This could benefit patients suffering from narcolepsy and other serious sleep disorders, but it also has applicability to shift workers, the military, and many other occupations where sleep is often limited, yet cognitive demand remains high."

What interested me most from Glenn's remarks at the conference was that the cutting edge of research isn't focused on increasing the length of alertness, but instead on enhancing the speed and quality of sleep. In fact, researchers at Columbia University are experimenting with transcranial magnetic stimulation (TMS), the same technology used by some neurotheologians to induce religious experiences, as a means of erasing fatigue. A field-portable TMS unit is under development.

Best practices now dictate that warrior-athletes stay out in the field for forty-eight hours straight, are brought in to sleep for sixteen hours, then head out for another forty-eight. Treatments to mitigate the need for sleep are as potentially helpful for commandos as they are for poker-table jockeys, but they still present the possibility of long-term negatives. However, imagine a drug that could quickly induce sleep so deep and restful that two or three hours produced a full recharge. Combine that with modafinil, or some future improvement of the same kind of drug. You would have warrior-athletes who could fight for two consecutive days and nights, grab all the regenerative sleep they needed in as much time as it takes to watch a long movie, then attack their enemies again, at peak energy. You would have the most feared and relentless fighting force in the world.

The committee's report was released in the middle of August 2008, under the title "Emergent Cognitive Neuroscience and Related Technologies." One of the most fascinating topics were pharmacological land mines. Instead of blowing someone up, these land mines would douse an enemy with brain-altering chemicals. Imagine a squad advancing, hoping to attack by surprise but instead getting instantly sedated, unable to function, possibly even obeying commands given by the people they had intended to kill. Meanwhile, subjects being interrogated in the future could receive harmless electrical pulses that interfere with their ability to lie. And the use of machines that will interface with humans to create super-performers is "limited only by imagination," according to the report. There's much more detail about these planned interfaces in the pages just ahead.

Not long after the meeting where Glenn, Chatterjee, and I spoke, the January 2008 issue of *Aviation Week* carried an article based on interviews with the woman who oversees some of the most provocative current DARPA research.

Dr. Amy Kruse began serving as a tech consultant to DARPA's director immediately after receiving her doctorate in neuroscience from the University of

Illinois. Now she is in charge of some of the agency's literally most mind-boggling research.

Among the items now on her plate: research into computer analysis of brain waves detected from satellites without the subjects' knowledge. It's expected to help intelligence analysts precisely identify and locate targets based on the hostile thoughts of enemy forces, and to help leaders in the theater of operations know if their deployed troops are alert enough to realize the intensity of whatever situations they're about to face.

As any veteran of Afghanistan or Iraq combat could tell you, it gets unbelievably hairy on the battlefield sometimes: incoming fire, orders being shouted, buddies taking hits, weapons jamming, and the likelihood of IEDs and/or ambushes along any efficient escape route. In such times of superstress, a combatant's mind can get so overwhelmed that it will lock in on just one aspect of all the stimuli. It may not even take in what the commander is trying to communicate.

Experiments have been run with soldiers whose brain waves are seen by their commanders via wireless computers. When a commander knows a soldier has gone into tunnel vision from information overload, he will know that he has to count on someone else for key actions and commands during an attack.

Here's how Kruse described her work at the annual DARPATech conference of 2005: "The operational environment will continue to become more crowded with information, so it is clear that our war fighters must be able to manage complex situations with faster, with more accurate and more concentrated cognitive capabilities. This means that issues such as cognitive overload, fatigue and decision-making under stress are fast becoming crucial factors in performance."

An earlier project called Augmented Cognition, AugCog for short, gave birth to something Kruse is working on currently. That is the Neurotechnology for Intelligence Analysts (NIA) program.

Under a $4 million, multiphase contract, Honeywell Aerospace, a major contractor for both the NIA and AugCog, has been developing the Honeywell Image Triage System (HITS) for DARPA. HITS divides satellite imagery into smaller image "chips" that intelligence analysts can view at a rate of five to twenty images per second.

Electrodes on the analyst's scalp pick up the increased brain activity that shows when she comes across a meaningful correlation while scanning rapidly. The analyst doesn't have to stop and think, form sentences, fill out a report, or even consciously register what may have been noticed only on a subconscious level. Her brain reports directly to the computer, via what is called a "man/machine interface."

Honeywell says this will allow analysts to process satellite data five to seven times faster. This boost is necessary for handling the high volume of imagery coming in, and turning those images into actionable information as quickly as possible. As is typical of neuroscience breakthroughs, development of the HITS system required blending the knowledge of many different disciplines, including psychology, electrical engineering, mechanical engineering, and avionics.

Phase Two of the NIA project began recently. Honeywell's work is being augmented with participation from Teledyne Scientific Imaging and from Columbia University, which is aggressively building its neuroscience reputation.

Phase Three is expected to produce a prototype that intelligence agencies will road test. According to Honeywell, the technology is nearly ready for operational use.

Altogether, AugCog and NIA research has linked several corporate and academic teams with the four different military services: DaimlerChrysler with the U.S. Marine Corps, Lockheed Martin with the Navy, Boeing with the Air Force, and Honeywell's team of some eleven industry and university partners with the Army.

Somewhat similar research into computerized monitoring of the awareness states of pilots has been under way for over a decade. Sensors inside a pilot's flight helmet pick up on the amplitude of his brain waves—basically the same technology used for neurofeedback relaxation studies. The computer, discovering that a pilot is becoming too fatigued, will initiate a cascade of events. First, the instrument panel will get brighter. Then it might start flashing on and off at such a rapid rate that the pilot's brain will respond subconsciously. Then it may start turning more and more functions over to an autopilot system.

According to Kruse, the Defense Department and all of the armed services are increasingly asking what else neurotechnology can possibly give them in the future.

Which is exactly why the CMIMEN&C/NSRNTD was created. One of the most brain-twisting and talked-about areas of recent research, for instance, is something called the Active Denial System, or ADS, which was demonstrated in late January 2007 at Moody Air Force Base in Georgia. ADS is a nonlethal but absolutely terrifying weapon system that can be field-mounted on a vehicle and will pulse electromagnetic radiation at target subjects as far as about five and a half football fields away. It instantly heats the water molecules under a target's skin to 130 degrees Fahrenheit. The effect is so painful, and so immediate, that it disrupts whatever the subject was trying to do, and inspires him, her, or all of them (the initially planned use for ADS is crowd control) to run, dive, flop, or do whatever else it takes to get out of the beam's path.

Even more ominous, ADS is the by-product of larger, ongoing research looking for technology that could delete and then replace a person's memories. That last sentence is guaranteed to make any true science-fiction fan remember at least three recent mainstream movies with variations on the same theme.

Another on-the-horizon reality actually seems like a direct lift from science fiction, specifically from *Minority Report*. In that hugely successful 2002 Steven Spielberg film, a year 2054 version of Tom Cruise is a cop who works with psychics. *Minority Report*'s "pre-cogs" (named for their talent, which is called precognition) were born to drug-addicted women and thereby have a useful mutation. They receive visions about murders before they are committed, including the killer's name, the victim's name, and assorted visual clues. Turning up in those visions gets a person arrested and stuck in an eerie penitentiary, even if the person is apprehended before the crime actually takes place.

And, in science fact, a team of neuroscientists has actually developed a brain scan–based way of finding hints about what a subject is intending to do. The scientists built their research on findings from earlier studies that used fMRI to understand what goes on in the human brain when the topic at hand is racial prejudice, violence, or lying.

It's a huge step from that knowledge base to actually translating brain activity patterns into discernible thoughts, and yet another big leap to reach official adoption and application of the process. But anyone looking at human nature from the dark side—which has been very easy to find in current headlines—can immediately think of dozens or more potential uses and abuses of precognition. Researchers I've spoken to think it may be a reality within thirty years.

Or perhaps a few years closer than that. In the March 6, 2008, edition of *Nature*, scientists from the University of California, Berkeley, reported development of a method for decoding patterns in visual areas of the brain, and using them to know what a subject is looking at.[2]

The scary potentials of this announcement are easy to conjure, but the researchers are focused on benevolent uses, like understanding the differences between various people in perceiving the seen world, and possibly accessing imagined visual content, like fantasies and dreams, perhaps to aid in psychotherapy.

By exposing subjects to visual images, and then recording brain activity, the UC Berkeley researchers first figured out a number of activation patterns. Then they built a mathematical model, an algorithm. Information from the patterns makes it possible for the researchers to look at brain activation and then make a well-informed guess about what kind of visual information has caused it to happen.

It was a very small study, just two subjects, both of them members of the research team. They scanned a set of 120 different images that they'd never seen before—mostly everyday objects like animals, houses, people, and so on. The computer was successful 110 times in figuring out what the subjects were looking at. When the number of images was upped to 1,000, the computer's success rate declined to 80 percent, a noticeable drop but still a high rate of success. "This indicates," the researchers wrote, "that fMRI signals contain a considerable amount of stimulus information and that this information can be successfully decoded in practice."

Asked for comment, researcher John-Dylan Haynes, a professor at the Bernstein Center for Computational Neuroscience Berlin and the Max Planck Institute for Human Cognitive and Brain Sciences, said the method worked out by the UC Berkeley researchers can decode only data that can be mapped out in space. This limits it to sensory inputs and body movements performed by subjects. A far more complex mathematical model would be needed for memories, emotions, or intended actions. However, the publication of the research in *Nature* is bound to make DARPA researchers and others in the field feel a significant step closer to that goal.

Even before 9/11, DARPA policy makers began focusing on small, rapid-strike forces to counter elusive enemies like Al Qaeda and the Taliban, transnational aggregations who can run a sortie, then recede into the local population or into difficult terrain. Much of DARPA's current research is being carried out with biologists whose aim is to make American forces not only swifter and stronger but also more resilient when up against fatigue, harsh conditions, and battlefield wounds. This emphasis on biological sciences intensified on June 18, 2001, when DARPA's direction was taken over by Tony Tether, an electrical engineer who received his doctorate at Stanford and went on to become founder and CEO of the Sequoia Group. His résumé also includes a stint as CEO of Dynamics Technology and as vice president of the Science Applications International Corporation (SAIC) Advanced Technology Sector, preceded by executive posts with Ford Aerospace, DARPA itself, and the National Intelligence Office in the Office of the Secretary of Defense.

Tether saw great potential in building better humans, and quickly expanded on DARPA efforts in that vein, primarily through a DARPA entity called the Defense Sciences Office. As reported by Noah Shachtman in *Wired*, DARPA petitioned Congress for $78 million in additional annual funding in early 2002, earmarked for research into "the development of biochemical materials for enhancement of performance."[3] Soon after, a DARPA document was drafted calling human beings "the weakest link" in defense systems, a situation that demands "sustaining and augmenting human performance" and "enabling new human capabilities."

In fact, DARPA announced its 2006 funding initiative under the title "Applications of Biology to Defense Applications." Although, as you would expect, it isn't easy for civilians to know everything the agency is up to, analysts believe that a majority of the programs are related in one way or another to neuroscience.

Of course, the trench warfare known as politics can pose a threat to the scientific advances sought to increase our fighting ability. Under the rule of former president George W. Bush, who thought "the jury is out" on the theory of evolution, scientists had to be on the lookout for faith-based decision making. Bush created the President's Council on Bioethics in November 2001. Its primary function turned out to be slowing down embryonic stem cell research, but its members also expressed reservations about research into augmentation of human capabilities. On February 27, 2004, Bush removed two prominent scientists, the cell biologist Dr. Elizabeth Blackburn and the medical ethicist Dr. William May, from the commission, ostensibly because they frequently differed with administration policy—like most contemporary scientists do—on stem cell research. Both were replaced by more "agreeable" folks, even though

Nobel laureate Thomas Cech, president of the Howard Hughes Medical Institute, calls Blackburn "a very smart and successful scientist . . . one of the top biomedical researchers in the world."

Even the Surviving Blood Loss program nearly became a casualty of science-fearing politicians, and was in limbo until Michael Goldblatt, head of the Defense Sciences Office, managed to get it reinstated. It's hard to imagine any argument against this program. Its intent is to keep massively injured warrior-athletes alive as long as possible, until they can be evacuated from the battlefield to get comprehensive treatment. Cues are being sought from nature—studying the ways that various organisms can partially shut down life functions while waiting out adverse conditions, like freezing weather. The Surviving Blood Loss program studies are aiming to make soldiers able to regulate metabolism on demand, entering a sort of hibernation. Civilian applications could follow, such as natural-disaster response and emergency medical treatment of severe wounds.

Another fascinating DARPA project involves a device to regulate body temperature. It has long been taught that muscles fatigue because vigorous movements cause a buildup of lactic acid. But in fact, muscles get tired because of excess heat. That is why our bodies release perspiration. Sweat is a means of reducing temperature. So is the Stanford-spawned invention called the Glove, which not only cools a person much faster than perspiration can, but also raises body temperature when that is what a situation requires.

Craig Heller and Dennis Grahn, biologists, started work on the Glove in the late 1990s. When an exhausted, severely overheated subject puts a hand in the device, a seal is secured around that person's wrist, and a vacuum pulls his blood to the surface of the hand. Within five minutes the subject's blood has cooled, and that person can go right back into action, reinvigorated. Someone about to enter hypothermic shock can, after two minutes in Gloves set for warming, feel himself rebalancing—even while remaining in a tank full of icy water. Stanford's football players wanted to experience the Glove after hearing how a lab technician who used to do one hundred pull-ups in his workouts had become able to do six hundred. In April 2003, to celebrate his own sixtieth birthday, Heller demonstrated one thousand push-ups.

The Glove was a bit clunky, something like an oversized tin can. Special Operations troops soon got the Glove in their own hands, and now they are testing newer editions, which are far more compact.

My own work of the past several years, trying to stay on top of who is doing what within all the areas of neuroscientific research, and where their work might lead, has introduced me to quite a few remarkable people. Jonathan Moreno is one of the most interesting of the entire group, and he has one of the highest vantage points for seeing the future of neurowarfare.

Moreno is a professor at the University of Pennsylvania and the editor of the journal *Science Progress*, as well as a senior fellow at the Center for American Progress. He holds a doctorate in philosophy, an unusual entrance ticket for the access he's gained into defense research.

Moreno's 2006 book, *Mind Wars*, begins with a scene that happened in 1962, when he was just ten years old. His father, a psychiatrist with a twenty-acre sanatorium in the Hudson Valley, was known for cutting-edge therapies. One day a yellow school bus pulled into the grounds with more than twenty young men and women aboard. Moreno enlisted them as playmates, putting together a softball game that lasted until the new arrivals had to go in and begin some sort of work. He didn't learn until the middle of his college years that his playmates had been bused there to take LSD, as prescribed by their psychiatrists in Manhattan. His father was one of the few physicians officially licensed to explore the therapeutic possibilities of LSD, marijuana, and cocaine.

Other LSD studies were going on at the Esalen Institute in Big Sur, and even though they ended abruptly, some of the preliminary findings gave Richard Tarnas impetus to write his books *The Passions of the Western Mind* and *Cosmos and Psyche*.

Of course, by the time Moreno learned what those people in the school bus had come for, LSD had become infamous. Goings-on at Timothy Leary's estate in Millbrook, New York, not far away, were among the many reasons why.

In 1994, Moreno was asked to join a presidential advisory committee looking into secret experiments that the American government had begun conducting in the 1940s, testing what happened when people were exposed to extreme radiation. Injections of plutonium were one method, though test subjects weren't informed about what their bodies were about to receive. During that committee work Moreno learned the CIA had been behind much of the research into LSD. Gradually, over the next few years, it dawned on him that since neuroscience is "perhaps the fastest growing scientific field, both in terms of numbers of scientists and knowledge being gained," the U.S. government must be just as determined to tap that knowledge base now as it was in all its previous decades of clandestine research.

Moreno's 1999 book, *Undue Risk*, compiled all he could gather about secret experiments sanctioned in the name of defense. It also made him popular with a great many men and women who believe the government is doing nefarious things intended to take over their brains. He speaks to these people with compassion, but doesn't believe that any such research is actually taking place. Even so, he was duly alarmed when he came across a DARPA strategic plan, released in February 2003, that said, "The long-term Defense implications of finding ways to turn thoughts into acts, if it can be developed, are enormous: imagine U.S. warfighters that only need use the power of their thoughts to do things at great distances."

Moreno reports the Department of Defense has around $68 billion to use annually on scientific research and development. The slice of the Pentagon's "black operations" budget that is earmarked for research and development he estimates at $6 billion or more.

At the end of a 2001 Dana Foundation conference on neuroethics, held in San Francisco, Moreno found himself raising his hand in front of a hundred neuroscientists to ask, "How come no one here said anything about how this applies to national defense?" He left the conference determined to start learning about what was happening regarding possible applications of neuroscience in defense circles.

First he contacted friends who were neuroscientists. No one would speak on the record. They all either were receiving DARPA funding or very much wanted to. But a few were willing to speak off the record, and they helped him get a general sense of some directions being pursued.

Next he Googled "DARPA and Neuroscience." Thousands of pages appeared. (A 2008 search registered 152,000 hits.)

Many of those pages were RFPs, which stands for Requests for Proposals. These documents ask military contractors to say if they are able to build certain devices DARPA wants to have, and also ask how much time and funding they would need. By mentally translating these RFPs from military-speak to English, Moreno learned more about what DARPA was hoping to achieve.

After *Mind Wars* was published in 2006, the National Research Council appointed him to lead the CMIMEN&C/NSRNTD, the same committee whose members explained to me much of the research mentioned in this chapter.

Moreno jokes that he's learned to think of the intelligence community, called "IC" in bureaucratic shorthand, as being pronounced "I See." Describing what some of our country's most estimable neuroscientists have recently told the committee, he reflects, "If I were coming up with conjectures like these, people would say I was crazy."

Bear in mind the prosthetic brain part I'm about to describe isn't a plot device from a sci-fi writer. It, and other equally bizarre notions of what is just over the horizon, are brought to you by leading neuroscientists. That prosthetic brain part is a hippocampus that is in effect a hard drive capable of interfacing directly with the brain. It is now being worked on by Ted Berger at the University of Southern California and others. Berger's first step is to create a version for testing on the brains of rats. They are making progress on how to download information into a brain. A few scientists currently believe that uploading data out of a subject's hippocampus into a data-storage device may also be doable, but will take longer to work out.

Right now the major debate in the field is whether it will be best to implant the artificial hippocampus or mount it externally.

The military is hoping for instant-training applications, such as learning the basics of a new language instantaneously, or memorizing data, such as photos of most-desired enemy combatants or the faces of everyone known to have been in an enemy training camp.

Moreno is highly aware of the debates now going on regarding these and other kinds of neuroscience-enabled human enhancements. We will look into that controversy more fully in the next chapter. But, according to Moreno, the enhancement debate will continue to shape up along these lines: "How far do we go, and who controls access to the technology and the information?" You've probably already realized that these are extremely large questions.

Moreno told me of a recent Department of Justice meeting in which participants talked about the possible future use of opioids as mass calming agents for crowd control. This discussion, of course, brought up images of the horrifically bungled and reprehensible *Nord-Ost* incident. The consensus in the meeting was that even if our technology reaches a point where knocking out troublesome people with synthetic opium would be practical, American society wouldn't want such weapons used. Of course, remembering Tiananmen Square as well as the *Nord-Ost* incident, we can't expect that all other societies would make the same decision.

This realization raises another tricky question: Should we develop the technology just so we can learn how to counteract it? Perhaps we really should—although if we do, we will increase the risk that the technology might get into the wrong hands.

Extreme stories and wildly paranoid suppositions about such weapons are frequently posted on the Web site of Mind Justice, which describes itself as "a human rights group working for the rights and protections of mental integrity and freedom from new technologies and weapons which target the mind and nervous system."

Moreno averages about one email per week from these people. A recent one claimed that former New York governor Eliot Spitzer was lured to his Mayflower Hotel hooker assignations by some form of external control.

I receive similar emails about once a month.

Moreno believes that the various neuroweapons already developed, and those on the horizon, would never be operated within our society, but only against enemy combatants. But he qualifies that belief as being just one guy's opinion, and adds that it may be human nature, at least for most of us, to want to deny that potential negatives may be looming. "We already operate in comfortable self-denial," he says, "regarding how much is already known and shared about us from our credit card usage, Internet clicks, and so on. So it's not hard to understand why there are so many paranoiac ideas in circulation, even among articulate, educated, and highly qualified people. And it is effectively impossible to disprove them." He has had several conversations with Cheryl Welsh, director of Mind Justice. Once he asked if there was any evidence he could present that might prove to her that she was wrong. Her response: "There isn't any way."

Moreno believes that robot armies are very likely to fight wars of the future. All the research into brain/machine interactions, directed by DARPA and others, is setting the technological stage. One of the first likely applications would be drone attack planes controlled by someone in a bunker whose brain wave activity directs the raining down of ordnance on enemy positions hidden miles away.

In early 2008, an article in *Wired* revealed powerful disagreements between Secretary of Defense Robert Gates and top Air Force brass about the deployment of Predator aircraft, and exactly how dollars earmarked for robotic bombers should be spent. If they're now arguing about such details, the robotic bombers are probably not far (figuratively and literally) from the horizon.

Information technology gave the world's dominant powers a significant military edge for decades. Neurotechnology will once again shift the balance of power on the battlefield back to countries with the substantial resources required for advancing its development.

This technological edge should curtail stateless enemies and rogue states. Unless, that is, faster than neurotech can fulfill its military promise, those enemies fully acquire the capacity to develop and deploy nuclear weapons. Useful technology spreads. That is a core truth. So we are now in an arms race to create the next generation of unimaginably potent weapons, and also in a race to contain the current generation of destructive capacity. That generation may seem outdated, but it's still the most convincing display of instant human-created hell we've ever seen. We have to realize, too, the spread of information technology is what has made asymmetric warfare possible. As we watch small groups of crafty individuals inflicting large amounts of damage to large, well-organized nation-states, we may richly enjoy revenge fantasies, imagining the new developments that will press our current enemies to the wall. But if we do achieve an era of relative peace, we will need to use that limited time with uncommon wisdom. Otherwise, neurotechnology's powers will eventually seep into the hands of rogue states and groups bent on derailing global harmony.

Before that, the next twenty years will witness the rise of performance enhanced warrior-athletes equipped with intention-scanning technologies designed to isolate and inoculate potential threats. The operative word here is potential. Imagine an advanced version of Honeywell's Image Triage System. It includes group behavioral-analysis software and individual emotion-recognition algorithms that key off micro emotional tics we all exhibit. Intelligence analysts and warrior-athletes will scour real-time surveillance feeds from satellites, unmanned predators, robotic insects, and other ingenious surveillance systems in their efforts to seek and destroy enemy combatants.

Because of the extraordinary personal, social, and economic cost of brain-related injuries, warrior-athletes headed for elite combat units will soon be genetically screened for a wide set of traits including mental resilience and physical stamina. Those who do not naturally meet these requirements will be augmented. To accelerate their learning prior to deployment, soldiers will leverage cogniceuticals and advanced cognitive-focused neural prosthetics to improve memory and attention. Sleep enablers will be widely used to optimize short-term performance and physical recovery between operations.

During forward operations the troops will be equipped with a host of drugs, administered via advanced drug device delivery systems, to improve their cognitive clarity, extend their physical stamina, and truncate the specific emotional trauma inherent in face-to-face combat. Postoperation periods will include virtual reality "come-down" environments. These will be used in conjunction with emoticeuticals designed to mute and reshape battlefield emotions, a highly refined "reframing" process with roots in today's cognitive behavioral therapy. Transcranial electromagnetic devices, variations on the "God helmet," will partially zap away events that could cause PTSD.

Despite the existence of chemical-weapons treaties, we will eventually see the rise of neuroweapons aimed at shifting the emotional and cognitive capacity of individuals and small populations. Memory bombs that give individuals short-term amnesia or electronic sleep-inducing weapons may seem in the realm of science fiction. But before the advent of atomic weapons so was the idea that 140,000 inhabitants of Hiroshima, Japan, could be wiped off the face of the planet with a single bomb. Neuroweapons will be intensely scrutinized by the world community, and we will all debate their enormous ethical and moral complexities. But history shows that humanity hasn't yet proved itself able to agree on plans for limiting technological developments that could manifest doomsday. Without a major shift in thinking, we won't limit the spread of neurowar capacities either. Governments and groups across the planet will increasingly tap neuroscientific advances in medicine, finance, marketing, law, and a host of other areas in order to gain military advantage.

Just days before our interview, Moreno was visited by a contingent of five Japanese scientists. They wanted to explore with him how they might build on the hope expressed at the end of *Mind Wars*, in the reflective final sentence that came to Moreno just before the manuscript was due at his publisher's office: "Perhaps, understanding more about this excruciatingly complex system, we can turn ourselves from the wars of the mind to the peace of the soul." If we're nearly as smart as our advancing neurotechnology suggests we are, we will find workable answers for this gently phrased challenge. It is my hope that neurotechnology will enable us to reach the benchmark Moreno imagined.

NINE

PERCEPTION SHIFT

Every fact depends for its value on how much we already know.

—Ralph Waldo Emerson

Reality leaves a lot to the imagination. —John Lennon

The TV producers wanted a sensational show. Anjan Chatterjee held firm. He lost out on tremendous free publicity, and even an Ivy League academic has much to gain if his work becomes better known to the world. We all seem to work in show business, one way or another, and fame can be more seductive than money itself.

Chatterjee, an associate professor at the University of Pennsylvania's Center for Cognitive Neuroscience, is at the forefront of a big story. It's going to keep turning within the world's consciousness—absolutely literally—for many decades to come. He wants that story told with the nuances left in, because they are extremely important. If this story gets reduced to sound bites for chowderheads, some of its unavoidable negatives will get magnified way out of proportion. Worse, essential progress in matters that mean the world to Chatterjee could get sidetracked or even snuffed, the way research into psychoactive drugs hit a stone wall because of rampant recreational drug use.

The TV producers lost interest, saying the story had to be simplified in order to be "palatable" for viewers, but quite a few other journalists and authors interviewed and wrote about Chatterjee after his article in the journal *Neurology* opened for public discussion the topic he calls "cosmetic neurology."

Chatterjee is a friendly and quietly engaging guy whom I've met at several conferences. Slightly built, young, he wears his deeply black beard trimmed close. He wouldn't seem out of place in a Bollywood romance, perhaps as the shy but earnest suitor who appears in act 2. But in fact Chatterjee is a highly respected, widely quoted scientist—a founding member of the Neuroethics Society, and on the editorial board of six academic journals including the *Journal of Cognitive Neuroscience*, *Cognitive Neuropsychology*, and *Policy Studies in Ethics, Law, and Technology*. He studied philosophy as an undergraduate and got his M.D. at Penn in 1985, followed by graduate fellowships and a faculty position at the University of Alabama in Birmingham, before joining Penn's stellar neurology researchers. His studies seek an understanding of cognitive systems, and extend into neuroesthetics as well, though his greatest focus may be in finding better treatment strategies for people trying to overcome brain damage.

In 2007 Chatterjee published an especially provocative paper in the *Cambridge Quarterly of Healthcare Ethics* in which he elaborated on parallels he sees between today's evidence of a cosmetic-neurology trend and the earlier development, and now widespread acceptance, of cosmetic surgery.[1]

Chatterjee projects that cosmetic neurology has begun, and will keep on, following a path similar to that of cosmetic surgery. Reconstructive surgery goes back at least 2,600 years, so it's based on deep-rooted human impulses. It took relatively recent technological shifts to allow cosmetic surgery to flourish in our time. Before anesthesia and antibiotics were developed, pain and the risk of infections limited the use of surgeries to change physical appearance. Those advances made cosmetic surgery more practical and ethical.

At the same time that reconstructive surgery advanced, America was becoming more urbanized. Most people's lives began to include more face-to-face contact with increasingly bigger numbers of co-workers and strangers. "Making a good impression" became a frequent necessity. Every aspect of personal appearance, and its power to win friends and influence people, got magnified. Even though plastic surgery was initially seen as an indulgence of vanity, it also held the promise of making people feel more self-confident in social situations.

The pioneering psychiatrist Alfred Adler, a colleague of Freud and Jung, and with those giants a co-founder of depth psychology, had a strong public impact in the 1920s and 1930s through his concept of the inferiority complex. Adler asserted that someone with negative inner feelings is likely to overcompensate, turning to behaviors that may be self-harming in the long run. More and more, people began to wonder if enhancing their looks could eliminate a source of inferiority feelings. It was a thought that fit like a dovetail joint with American notions of self-improvement and social mobility.

In fact, experiments were done in the 1920s and 1930s with reshaping the faces of convicted felons, to see if their psyches would make a parallel positive shift. Then, in the years that followed, the American ideal of beauty became even more powerfully linked with youth. So it became more common for people to try to keep themselves looking young. The face-lift procedure was developed, along with chemical peels, dermabrasion, tummy tucks, and other ways of trying to appear more attractive. Smaller breasts for men and bigger ones for women, rib removals to accentuate an hourglass shape, removing fat cells through liposuction, implants for prominent chins and cheekbones, and numerous other ways of tweaking external appearance have become well known and widely practiced. In 2006 board-certified surgeons performed nearly 10 million cosmetic surgical procedures, seven times as many as were done in 1994.

In effect, people found a need for cosmetic surgery based on their inner feelings. They sensed a potential for better inner feelings, and connected to the idea that a physical transformation would win that result.

Cosmetic neurology aims straight at the source, promising better inner feelings through technology-enabled neurological transformation. The avalanche of psychotropic medicines coming onto the market in the last couple of decades is one of many clear signals that quite a large number of people now realize that they could feel a great deal better psychologically. Baby boomers, having seen their parents age, are acutely aware that growing old often brings dementia, and that old age is what they are moving closer to every day. In fact, it's estimated that between one-fourth and one-half of us will experience dementia by age eighty-five.

Among younger Americans, emotional and attention disorders are now seen as almost epidemic. Part of this may be that, given some good tools for interventions, we're starting to pay attention to conditions that we generally ignored when we didn't know what caused them or how to fix them. Part of it may be the accelerating pace of change, possibly leading to a great sense of helplessness and dislocation. Whatever the reason, several hundred companies are now working to perfect neurotechnologies that will be light-years improved from current offerings, more precisely targeted, less burdened by side effects, more user-friendly overall. There's no question in my mind people are going to embrace these pharmaceuticals, and in time neurodevices, with the hope of improving their lives.

There's an exquisite sign of how widespread cosmetic neurology will become in a friendly sounding word—"pharmies." That's a popular slang term on many high school and college campuses, and among legions of men and women in the early phases of their careers. It refers to the prescription medicines, the pharmaceuticals, so many young people stockpile for their own use, or to trade or just give to any friend who has a problem that the pills might resolve. Many senior citizens have been doing the same thing for quite a long time. The Drug Enforcement Administration has a little too much on its hands with cocaine and marijuana smugglers, not to mention crack houses and meth labs, to track and prosecute the very elderly and the very young for these unlawful transactions.

In a 2005 survey of more than ten thousand college students it was reported that between 4 and 7 percent of them had tried attention-deficit-disorder drugs for either all-night cramming sessions or to do better on their exams. On some campuses more than 25 percent of students had used the pills.

Informal research suggests the practice is now even more widespread. After a recent talk at an Ivy League college, I spoke with a professor who had performed blind class surveys on the use of stimulants in undergraduate classes. She said 60 percent and more of her students reported using them at some time, while about 80 percent knew someone else who used them. Perhaps the differences can be traced to a higher level of competitive pressure among Ivy League students, but I think it's more about an accelerating surge in pharmie consumption, on campuses everywhere, since the 2005 survey.

Despite Nancy Reagan's former "Just say no to drugs" campaign, and despite the potentially life-threatening side effects of pharmies, I expect that surge to continue, especially after treatments with fewer side effects reach the market.

Neuroscientists are working on technology now that will revolutionize the design and testing of future pharmaceuticals, lopping years of testing and millions of dollars off the development cycle.

Pharmies come from Web sites that sell them illegally, without prescription, or perhaps on presentation of faked scrip. They also come from legitimate sources, sometimes for real ailments and sometimes for well-rehearsed descriptions of the right symptoms. However the pills are secured, they can relax a tense person, bring on alertness when you're drowsy, lift the fog of depression, get you back on track after too much marijuana or alcohol, and stanch physical and emotional pain. The Internet provides plenty of research materials about how the drugs work, and their potential side effects, and being your own pharmacist and prescribing doctor is supremely faster and cheaper than going by the legally established system.

It can be dangerous as hell too, of course. The sensationalism that Chatterjee would like to hold at bay won't stay damped down forever. Witness the unfortunate death of Heath Ledger, the Oscar-nominated star of the 2005 film *Brokeback Mountain* and an actor who showed even more of his tremendous promise as the psychotic Joker in *The Dark Knight*. The *New York Times* compared him to James Dean, both in terms of the tragedy of his death and in terms of his impactful acting. A journalist in his native Australia wrote, "His extraordinary version of a gay cowboy . . . was impossibly beautiful and right at the fringes of human expression."

Suffering from anxiety, insomnia, pneumonia, and depression, the twenty-eight-year-old Ledger reportedly had six different prescription medicines coursing through his veins when he died: oxycodone, hydrocodone, diazepam, temazepam, alprazolam, and doxylamine. Some of these drugs can kick up the effects of others, sort of like wind chill increases the bite of cold weather. Taking diazepam or temazepam with oxycodone, for example, boosts that drug's already powerful opioid effects. Oxycodone, sometimes marketed as OxyContin, is a synthetic drug that is chemically close to codeine.

There are several routes medications can use to metabolize. When medications compete for that same route, problems can occur. One possibility is that modifying your intake of one drug, whether upward or downward, may intensify what the other drugs in your system are doing. With powerful medications at play, lethal results are easy to trigger.

The death of a young, handsome, gifted actor is a dramatic cultural event.

People still make pilgrimages to Cholame, California, which lies between Paso Robles and the Pacific Ocean, to see the intersection where James Dean died in 1955 after he crashed his Porsche Spyder. But this cultural event in the making, signified now by cosmetic neurology, is going to be more overwhelming than all tragic movie deaths put together.

Here's how I look at it, in neurosociety terms.

We live in a highly connected, urbanized world, the result of thousands of years that we humans have spent trying to control our physical environment. Our sociocultural world has evolved slowly over the past several thousand years, but much more rapidly in recent time. It has moved light-years faster than the evolution of our physical beings and our brains in particular. The truth is that our relatively primitive brains can and do get severely stressed by our modern world. Just using the word "stress" tips off the situation. It's a term originally used by metallurgists to describe what happens to metal that gets tortured until it threatens to rip. We've had to adopt that word to describe adequately how our brains sometimes feel.

So as science gives us more tools to control our mental and emotional environment, we eagerly take them up and use them. The SSRI drugs discussed by Stanford's Brian Knutson were developed for the treatment of depression, but soon proved themselves helpful for quite a few different maladies, including anxiety and insomnia. A common denominator among illnesses treatable with SSRI medications is that most are caused by stress. It's no surprise, then, SSRIs are the world's most frequently prescribed psychopharmaceutical.

Though they've helped millions, SSRIs come bundled with some terrible potential drawbacks. Side effects can include nausea, suicidal thoughts, and loss of sex drive. Many people report hellish symptoms when withdrawing from SSRIs. As Iowa Senator Tom Harkin recently remarked to me at a breakfast, recalling the experiences related by some of his constituents, "They can be harder to get off than crack." So they're not an optimum tool. But they are, along with variants that influence levels of other important neurotransmitters like dopamine and norepinephrine, probably the best medicines we currently have. And we're using them (pardon the expression) like crazy.

SSRIs, among all prescription drugs, may have done the most to create the conditions that have allowed this extremely casual acceptance of playing pharmacist for one's peers. They brought into public discussion the fact that our brain chemistry is in charge of our mental states, and that it can be manipulated. Today's students may already have been taking such drugs for years to battle depression, anxiety, or attention deficit disorder. If they haven't, they know plenty of peers who have. A recent *New England Journal of Medicine* report said that antidepressants are now prescribed to some 50 percent of the students seen at college health centers, including students who have schooled themselves to present convincing symptoms for attention deficit disorder or depression, and employ that skill set to take home drugs they will use in search of a competitive advantage.

Today's students have also grown up in an age when drug companies can advertise directly to consumers, a result of a 1997 Food and Drug Administration decision that has made rapidly read lists of side effects a staple of late-night ads, and also of comedians' routines. These ads can be laughable, but they also further a potentially dangerous idea: You can take a pill for just about anything you want to fix.

Meanwhile, people who scour the Internet for in-depth information on pharmaceuticals can sometimes feel that they know more than their doctors—whose acquisition of information could be skewed by offers from Big Pharma companies to support their research or payments to attend plush conferences, possibly paying for presentations. Doctors have reputations to protect in their communities. Internet sites thrive on liveliness, and may not be afraid to stand behind controversial approaches. Most patients have become aware that guesswork is going on, no matter whose advice they follow. Prescribing physicians, even the best-regarded psychopharmacologists, often have to grope to determine the right meds for a given patient, trying different pills and different combinations until they discover what works best with that particular person's biochemistry.

Phoenix House, which describes itself as the nation's leading nonprofit provider of substance-abuse treatment and prevention services, with over one hundred programs in nine states, has shifted emphasis on its youth drug-prevention Web site, Facts on Tap, to include materials on abuse of prescription drugs. The site is administered by the American Council for Drug Education and the Children of Alcoholics Foundation, which are Phoenix House affiliates.

With all that officialdom behind it, the Facts on Tap site is very symmetrically designed, responsible looking, and more than a bit stuffy. There's more fun to be had on CrazyBoards, where freewheeling discussion boards on topics like depression are moderated by people using such names as Velvet Elvis,

HaloGirl 66, and Greenyflower. If you want to discuss obsessive-compulsive disorder, you're instructed to "Click Here—Repeatedly."

That irreverence is a good sign younger people are the chief demographic. They tend to be veterans of social networking sites like MySpace and Facebook, and probably have fewer inhibitions than earlier generations about laying out extremely personal data. Despite its being a breach of CrazyBoards protocol, members often casually mention how many half-empty bottles of prescription meds they have lying around. From that point, it's easy enough for pharmie seekers to go off-site and exchange e-mails and more.

Inevitably, some of these exchanges will lead to regret. But even more inevitably, they will keep on taking place. Some blame must be placed on the high costs of health care in America. There's a parallel situation in the trend of Americans taking dental and medical "vacations" to India, Romania, Thailand, Mexico, and other destinations in order to have procedures done that they couldn't afford in their own country. Some travel agencies specialize in helping to organize these trips. They can direct customers to recommendations and testimonials for particular dentists, doctors, and surgeons. While it takes a leap of faith to entrust your teeth to someone you've never met who lives in a place you've never visited before and was trained someplace you've never heard of, people like having control over their lives, and it's inevitable that they will try unorthodox channels to reach their goal—whether it's excision of a tumor or recovery from a pot binge in time for midterm exams.

Use of other neuropharmaceuticals, including prescription stimulants, is skyrocketing. For example, the prescribing of drugs to treat attention deficit/hyperactivity disorder in adults aged twenty to thirty more than tripled in the United States from 2000 to 2007. Some 14 percent of students at a midwestern liberal arts college reported borrowing or buying ADHD medications from one another, and 44 percent reported they knew of someone who did. The students say they use these new tools to feel better—less depressed, less stressed out, more focused, better rested. In other words, to gain control of their mind.

The world's seventy thousand neuroscience researchers are busy attempting to decode the neurobiology of nearly every imaginable emotional, sensory, and cognitive state. With their progress propelled by the latest techniques in genetics and brain imaging, neuroscientists expect to create more targeted therapies that can more safely treat a wide range of neurological and psychiatric illnesses. Moreover, just as today's drugs are being used by healthy individuals to satisfy a whole host of desires, tomorrow's more sophisticated tools will enable individuals to ever more precisely influence their own neurochemistry. This newfound ability will lead to something very big—a fundamental shift in each individual's perception of daily events, ultimately transforming personal relationships, political opinions, and cultural beliefs all around the world. Revamping our minds, being freed from old constraints, we will literally and figuratively see the world in a brand-new way.

Futurists call this coming reality "neuroenhancement." I prefer a slightly different term, "neuroenablement." I'll explain why a little further ahead.

Beyond Therapy: Biotechnology and the Pursuit of Happiness, a report released in 2004 by the President's Council on Bioethics, says we as a society must consider the deep morality of these issues, and cautions against using new technologies for enhancement purposes:

> We want better children—but not by turning procreation into manufacture or by altering their brains to give them an edge over their peers. We want to perform better in the activities of life—but not by becoming mere creatures of our chemists or by turning ourselves into tools designed to win and achieve in inhuman ways. We want longer lives—but not at the cost of living carelessly or shallowly with diminished aspiration for living well, and not by becoming people so obsessed with our longevity that we care little about the next generations. We want to be happy—but not because of a drug that gives us happy feelings without the real loves, attachments and achievements that are essential to true human flourishing.

The caution is worth sounding, but our government won't be able to stop this cultural change from happening. Like it or not, in spite of all the king's horses and all the king's men, the shaping of one's perception with neurotechnology is coming, and everyone should be prepared. Treatments meant for therapy are going to be used for personal enhancement. Or, as I like to phrase it, enablement. That's not sensationalism. It's a solid fact within a highly nuanced situation. It may not seem publicly palatable to politicians or TV executives yet, given the complexities, but being ignored by branches of government or media doesn't mean much: as Neil Young once wrote, "I can't tell them how to feel . . . Sooner or later, it all gets real."

The famous philosopher Wittgenstein once said, "The limits of our language are the limits of our world." Indeed, language expands with need. Back in the nineteenth century, long before there were SSRIs or any other effective treatments for depression and other forms of mental illness, only two words were used to describe mental disorders: "lunacy" and "feeblemindedness." As treatments advanced, so did the words used in diagnosis. The same will be true in the therapy-versus-enhancement discussions.

Establishing the line between therapy and enhancement is going to be a contentious and difficult project, ongoing for many years, rather like the street-ripping and building-demolishing phases of constructing mass-transit systems through long-established urban patchworks.

The long-established patchwork in this case is humanity's selection of diverse and frequently sacrosanct ideas about what is correct behavior for humans. The debate will revolve—as it already does among those people who know what is ahead on the highway of history—around a number of moral concerns, philosophical approaches, and practical safety issues as to the proper role of medicine, therapeutics, and values in different societies.

So, what is normal? Clearly, every single one of us enters the world with different natural endowments, emotional, cognitive, and sensory capabilities, which then develop at varying rates and to different degrees. As the tools of neurotechnology are used for personal improvement, cultural disagreements about what is "natural" will generate political and religious tensions. Do we, or do we not, have the basic right to augment ourselves? Some will argue that the capabilities of the human mind should not be limited. Another inevitable point of view is that access to enhancement technologies will surely be unequal, and this will make existing gaps of social inequality even greater. Some will want neural enhancement completely outlawed. Others on both ends of the age spectrum will get their pharmies by any means necessary and all avenues possible, just as they're already doing. LSD and Ecstasy enthusiasts have continued, even though their drugs of choice have been outlawed for years. Meanwhile, marijuana is still thought to be one of America's biggest cash crops, while cocaine is still popular enough to keep sections of South America and South Central Los Angeles, to name just a couple of affected areas, perpetually on the brink of instability. Aging baby boomers, the herd that has pushed so many trends along, aren't likely to lie down for restrictions on anything that will ease their coming old age, either.

In today's terminology the use of neurotechnologies by "healthy" individuals for "nonmedical" purposes is called "neuroenhancement." However, that word does not capture the actual intention and belief of many of us who are looking at what the near future holds. This is why I used the term "neuroenablement" for many of today's emerging technologies. "Enablement" implies lifting "the bottom up," and addresses issues of fairness and social inequity. "Neuroenablement" seems to me to be a term that empowers people: It connotes the idea that people will begin leveraging better tools for mental health to bring themselves to where they most want to be regarding emotional, cognitive, and sensory capacities, to a more optimal point along the normal distribution range.

When I hear people use "enhancement," I get the feeling that they're thinking some superhuman state. In that sense, most neurotechnologies in development that will become available over the next decade do not qualify.

The advances being pursued today are primarily neuroenablers. They will allow those of us who are in the lower to middle percentiles to perform in the high percentiles, relative to our best-performing peers. For example, there are no cognitive enhancers in development that would allow your intelligence to leap beyond Einstein's IQ. However, there are cogniceuticals coming down the pipeline that have been shown to improve the average intelligence of those in the study. While no studies yet exist that specifically test these drugs in people with the highest IQs, there is anecdotal evidence from the U.S. Air Force that natural peak performers don't gain an additional edge when they use the alertness enabler modafinil.

In a nod to Wittgenstein's remark about the limits of language, I like to think we will enjoy a saner discussion of neuroethics if we use "neuroenablement" as the word of choice. It may help regulators to develop effective policies with respect to the virtually unstoppable advances of neurotechnology, and in the end, it may possibly promote human dignity.

But words, for all their emotional freight, are just signposts. It's far more important to ask what is it we will want from enablement (or enhancement).

Cognition is high on the list. It would be great to learn concepts and skills faster, make more rational decisions, remember things more easily, stay better focused, just plain be smarter, especially as our physical bodies stay healthy longer. Today a Irvine, California, company has even given itself the name Cortex Pharmaceuticals, obviously hoping that baby boomers will soon happily pay to avoid having those lapses of recall that get chided with "I guess you're having a senior moment." Cortex Pharmaceuticals is just one of over forty companies now developing memory-improving compounds.

Do you want to play classical music as beautifully as Yo-Yo Ma, but the only open date at Carnegie Hall is in six months? You were just traded to the Los Angeles Lakers and need to master Phil Jackson's triangle offense overnight so Kobe Bryant won't get mad at you? Someday we may be able to get pretty close to quickly meeting such challenges.

Better control of our emotions will be an even greater prize. Nearly all of us could do with less anger, or with more skillful harnessing of appropriate anger, so it can help us reach resolution in situations where we have truly been wronged. More empathy, as discussed earlier regarding the work of Paul Zak, Mike McCullough, V. S. Ramachandran, and others, could make the world an infinitely better place. Meanwhile, given a better range of choice and better ability to make smart, constructive choices through neuroenablement, most of us are likely to opt for greater pleasure and less pain.

Our emotions are mostly triggered within the oldest parts of our brains. When we have them under a reasonable amount of control, emotions are extremely valuable responses to our environments. Shutting them down is not only a stupid idea, it saps much of life's potential pleasure. Neurotech will provide powerful tools for overcoming the constraints that crop up because our physical and social environments have evolved faster than our biology. It will corral mental illness and neurological disease to a great extent; it will subtract a major portion of the world's misery.

As a ground-level factor in cognition, we may soon enable our basic senses to be both finer tuned and broader in reach: more precise hearing, olfaction, and taste; more sensitive touch when we want it, in the bedroom and elsewhere, but less sensitivity when we have to endure extremes of heat or cold.

Our neurochemistry is influenced by a broad set of modulators with a dizzying array of combinations. For example, dopamine is important in reward and plasticity, the ability of the brain to adapt. Serotonin is key in aggression and mood. Opioids regulate pain and pleasure. And these basic building blocks interact with cultural advances, and when they are at appropriate levels for whatever situations we face, we are like beautifully voiced musical instruments that stay in tune, all tones and overtones working together. Neuroenablers will make it possible for more people to play along in the orchestra of life. We will have steadier emotional and behavioral pitch; in scientific terms, an extraordinary amount of behavioral flexibility.

Improve all of these things—sensation, cognition, and its sister, emotion—and you don't need to aspire to being superhuman. Just being fully human, consistently at your best, with all the positive faculties you need whenever you need them, will be blessing enough. And, while not taking my eyes off the prize of stronger empathy, let me add that being fully at peak capability will be the ultimate competitive advantage for any individual or organization—up to the nation-state level and beyond.

Will people choose the neuroenablement path to improve their mental performance? And will that choice result in greater progress?

I'm betting that both of those are going to happen, at dramatically high levels. Anyone who follows sports knows athletes in many different fields have used physical enhancers to compete and win, even illegally and at the risk of their careers. Anyone who tunes into entertainment events from movies and TV shows to concerts—for that matter, most anyone who takes a close look around his or her workplace—knows who people have chosen cosmetic surgery to shape their appearance. If ways are found to safely improve human capital productivity, workers will use these new tools to increase their efficiency and keep their jobs.

We see this already in how classical musicians deal with stage fright. The requirements for classical musicians include delivering a technically superb performance of complex music the audience already knows intimately, adding expressive nuances that lift their playing above ordinary perfection, and not getting rattled. The same predicament is nicely described in the song "Stage Fright" by The Band: "Just one more nightmare you can stand."

Stage fright is a paralyzing kind of hyperfocus, in which nearly anything unexpected can send your confidence in your ability to perform right into the tank. It's an example of a fight-or-flight response, where primitive parts of the brain torque the production of adrenaline in order to supply you with quick sustained energy to either run away or fight for your life. When you are highly adrenalized, your mind keeps pushing choices at you, and making you feel that they have life-or-death significance. Meanwhile, if you're a classical musician, there's an audience waiting for you to make fine choices and sublime music.

Beta-blockers, which were developed to counteract high blood pressure, help control stage fright by inhibiting adrenaline receptors throughout the body, brain included. The adrenaline may rush, but it can't take over your behavior quite so easily. As a result, an unofficially estimated 25 percent to 75 percent of classical musicians pop pills before demanding situations, usually either beta-blockers or medicines for attention deficit/hyperactivity disorder.

Even the doctors who prescribe beta-blockers to performers will often use them for their own stage-fright situations, such as needing to present a paper at a medical conference.

The downside, which depends on individual tolerance as well as amount and frequency of dosage, includes the chance of anxiety, headaches, insomnia, and appetite loss with the attention-deficit-disorder meds, and drowsiness with beta-blockers.

There are many fields where prescription drugs are semicovertly embraced in the interest of clear minds, improved concentration, and reined-in emotions. Do you recall Paul Phillips, the thirty-five-year-old former computer programmer who used Adderall and Provigil to help him pull in some $2.3 million from poker tournaments? Tournament poker is dominated by those who can play at a high level for the longest possible time. If you can imagine what it would be like to put in a sixty-hour workweek, think what it would be like if everyone you were working with was trying to slit your throat—metaphorically speaking.

Ritalin, the drug that has for years been prescribed to those who are diagnosed with ADHD, actually increases your mental activity, but also focuses it more narrowly on what is immediately in front of you. It's likely that every teacher in America has at least one student who is taking Ritalin to stay calm enough to pay attention in class. Long-haul truckers, who have profitable days when they can rack up as many hours as the law allows, have been fans of Ritalin longer than the educational establishment. Poker also demands a tight attentional focus, both to your own cards and also to whatever tics and tendencies you can discover among your competitors. If you were to use that extremely popular performance enabler found at espresso bars for a marathon poker tournament, you would run into that syndrome known as "coffee jitters." But Ritalin, and beta-blockers like Inderal, are known to give some people a relatively smooth and sustained effect.

The fact that Phillips gives psychopharmacology credit for building his bank account suggests drug companies will encourage their research departments to develop next-generation neuroenablers of various kinds as rapidly as possible. In the words of University of Pennsylvania bioethicist Paul Root Wolpe, "Whatever company comes out with the first memory pill is going to put Viagra to shame."

As more people live longer and global competition intensifies, people will need to learn new skills throughout their lives. Neuroenablers would make continual education feasible and perhaps, in time, less expensive. In this way neuroenablers represent the next form of competitive advantage beyond information technology, a neurocompetitive advantage.

Any big change naturally inspires protest. There will be divisions over whether or not we should embrace this new way of living. Every nation and culture and subculture is going to react differently. While the United States, Great Britain, Nazi Germany, and the late Soviet Union were all industrial powerhouses, each country chose to direct its technological capital toward different visions of the future. In a similar way, socially and legally approved uses of neurotechnology will vary widely between Singapore, India, China, and the United States. However, because we live in a highly competitive global economy, the rewards of being a neuroenabled powerhouse will be great. When even just a few individuals choose to improve their mental performance, their choice will transform the basis of business competition for the rest of us.

Neuroenablement will start a cascade. One outcome will be a radical transformation of today's prevailing view of managerial common sense about how to achieve highest productivity. It will alter cost structures, transforming both productivity and social relations in the workplace, schools, families, and all of our other social institutions. It will provide individuals, companies, countries, and societies with new tools to let us produce more work with less strain, less effort, a more lasting in-the-zone feeling.

Athletic competitions take place in what folklorists call a "magic circle."

That's a place where special rules apply, such as how long a game lasts or how hard you can legally bang bodies with your opponent while you both compete for a rebound. So there are lots of people in striped shirts, or in white lab coats, dedicated to seeing the rules of the games are followed. But in the marketplace, in the workplace, in schools, in poker games, on golf courses, and in orchestra pits, how can would-be high achievers be stopped from using legal medicines in order to stay at a mental peak state? The economy would have to support armies of neurocops. Furthermore, would we want to stop these people and their tall aspirations? Probably not, if the person taking those drugs was going to find you the best possible investments from among several complex choices, pull you through a high-risk surgery, or find all the shoe bombs before the terrorists got on the jumbo jet.

But there is more to consider in this cultural change. For instance, behavioral differences will emerge as individuals begin to shape consciously their emotions and thus their perceptions. How will their shifted perceptions change their decisions, and go on to change the world around us? Will we all be on competitive treadmills? Or will this higher productivity create life options in which people can choose to work less, yet still have a secure and pleasurable living standard? Will consistently upbeat emotions be the right basis for making decisions?

Every month the U.S. government conducts its Consumer Confidence Survey. It's a bit of intelligence gathering that helps the Federal Reserve decide where to set interest rates. Through the survey, the perceptions of ordinary people are asked about the world around them—particularly whether their current emotions about money are confident or fearful—play an influential role in the macroeconomic decisions of our government. As individuals tone down fear and anxiety with emoticeuticals, they will perceive differently. This will ultimately affect which monetary policies get implemented. What might happen when substantial shifts in specific basic emotions become possible? How will this impact how we *feel* about the events that define our lives? And what could be more important than feelings? For example, most of us will be drawn toward choosing confidence over fear. It will feel better. But will it always lead to good decisions? Sometimes it's wise to worry.

Nothing less than war and peace are driven by our aggregated perception of events. Art, marriage, birth, death, disease, and religion are all powerful spurs to human feelings. By making us feel uplifted and thankful, or destitute and hopeless, they drive our actions. To quote Harvard psychologist Danny Gilbert, "Feelings don't just matter, they are what mattering *means*."

Researchers who are trying to define the how and why of feelings are creating a cornerstone of the emerging neurosociety. A team of neuroscientists led by Nathalie Camille at France's Institute of Cognitive Science in Bron has been using brain imaging to research what happens when we must make decisions without clear knowledge of how they'll turn out. The researchers' most recent findings, published recently in *Science*, show that a decision making region in the brain, the orbitofrontal cortex, does much of the work involved in mediating the experience of regret.

"Facing the consequence of a decision we made can trigger emotions like satisfaction, relief, or regret, which reflect our assessment of what was gained compared to what would have been gained by making a different decision," Camille and her associates reported. The cognitive process is known as counterfactual thinking. That basically means the process of learning from past mistakes.

The researchers presented subjects with a simple gambling task, and recorded each subject's choices in terms of their anticipated and actual emotional impact. Normal subjects reported emotional responses that showed they were able to do counterfactual thinking. But some subjects were patients with lesions in their orbitofrontal cortices. They simply did not report any regret, or anticipate negative consequences that might come from their choices. This syncs up with reports that some convicted murderers, when autopsied after their executions, had significant damage to the frontal areas of their brain from previous physical traumas, such as head injuries in car crashes. It also raises the question of how controlling emotions one feels right now might affect decisions made tomorrow.

As neurotechnology advances and the precise neurobiology of regret is understood, it's possible that we will be able to choose how much regret we feel. It's impossible to guess how this might impact personal relationships or our perceptions of daily life. Will it allow people to be more heartless in how they treat others, like murderers who don't experience regret over taking lives? Or will it make people quicker to relinquish old grudges, to fortify their constructive impulse toward forgiveness and moving on?

Your guess is as good as anyone else's. But this much is certain: By providing new tools to influence human emotion, cognition, and sensory systems, neurotechnology will create profound consequences in how we perceive social, political, and cultural problems. That's why studying the societal implications of neurotechnology, and immediately, is so critical. We have more of a chance to be ready for this technological wave than humanity did in previous waves of societal transformations like the agricultural, industrial, and information revolutions.

Meanwhile, as more and more biology and chemistry labs focus on developing next-generation brain drugs, other researchers working on a different piece of the neurotechnology revolution are designing implantable medical devices that interact with the brain through tiny electrical impulses.

Implanting a medical device through a surgical procedure is clearly more complex, time consuming, and expensive than popping a pill, so the implications for neuroenhancement with implanted devices are less immediate. However, advancing nanotechnology-manufacturing techniques will eventually shrink the size of the implantable devices. Surgeries to implant them will become less invasive. Over the next two decades, the impact of neurodevices will be profound.

Many neurodevices are in broad use today. They include deep brain stimulators to reduce the tremors from Parkinson's disease, spinal cord stimulators to treat intractable pain, and cochlear implants for the deaf. Moreover, there are teams of doctors across the globe testing implanted devices to treat brain-related illnesses like Alzheimer's, depression, addiction, and obsessive-compulsive disorder, to name a few conditions that might soon be more easily managed.

More than one hundred thousand people worldwide are already neuroenabled by cochlear implants. Unlike a traditional hearing aid, which simply amplifies sound that enters the ear canal, cochlear implants transduce sound from an external microphone and relay the signal to electrode arrays that directly stimulate nerve fibers in the inner ear, bypassing the auditory sensory apparatus entirely. The devices work by splicing incoming sound into different frequency ranges that are then passed down separate wires contacting nerve fibers directly. In short, this amazing prosthetic enables the deaf to hear. One only needs to read Michael Chorost's entertaining book, *Rebuilt: How Becoming Part Computer Made Me More Human*, to understand how developments in cochlear implant technologies will lead to superhuman hearing. (Chorost originally wanted to call the book *Mike 2.0*.)

Retinal implants are still in clinical development, but are expected to reach the market in the next few years. The current goal of retinal implants is to enable patients to "see" objects by identifying their size, position, and movement, allowing a previously blind person to move independently in an unknown environment without the need for a guide dog or cane. There are currently 35 million people across the globe who are either blind or severely visually impaired, and who could benefit from this coming technology. By 2020, this figure is projected to double. As with the cochlear implants, it's easy to imagine this technology progressing enough to expand the visual capacity of humans.

Beyond the expansion of our sensory capabilities, these devices could also transform how we feel about the world around us. In *Looking for Spinoza*, Antonio Damasio details his theory that we experience a chain reaction begins whenever an emotion (which he defines as a change in body state in response to an external stimulus) triggers a feeling (the representation of change in the brain as well as specific mental images). Damasio believes that feelings do not cause bodily symptoms. The opposite happens: We do not tremble because we feel afraid; we feel afraid because we tremble. That's the direction in which cause and effect flow.

If Damasio is right, then by directly influencing our nervous systems, for example by reducing our body's reactivity to trembling, we will in fact influence our mind's conception of ourselves and our environment.

The cutting edge of neurodevices is an area of research around brain-computer interface (BCI). In one-way BCIs, computers either accept commands from the brain or send signals to it (for example, to restore vision). Two-way BCIs would allow brains and external devices to exchange information in both directions, but they haven't yet been successfully implanted in animals or humans. However, as you saw in the previous chapter, there is tremendous interest among the defense community in developing this capability, and plenty of money to help make it happen.

Nothing brings home the future impact of these devices more than watching a neurosurgeon implant an electrode in the brain of a severely depressed patient, and seeing the expression on the individual's face literally shift from a frown to a grinning smile in a matter of seconds, as the physician turns on and calibrates the amount of electrical stimulation to the brain region being targeted. Doctors who have performed this delicate procedure note in rare cases they have moved their patients to a state of orgasmic ecstasy. Clearly, Woody

Allen's "Orgasmatron," made famous in his 1973 film *Sleeper*, isn't as far-fetched as one would have imagined even a decade ago.

Looking even further into the future, some researchers believe we are entering a period of "paradise engineering." In his online manifesto, *The Hedonistic Imperative*, futurist David Pearce says we are poised to explore a spectrum of mental superhealth—emotional, intellectual, and ethical—which will soon become safely accessible to all, thanks to advances in genetics and neurotechnology. "Early in the 21st Century, the prospect of paradise-engineering still sounds weird, Brave New Worldish—and perhaps 'unnatural.' Yet the metabolic pathways underlying heavenly states of consciousness are neither more nor less 'natural' than any other patterns of matter and energy instantiated elsewhere in space-time."[2]

The coming neurosociety, then, will probably be populated by people with far fewer physical and psychological ailments. Traumatization that occurs in utero and in early childhood may no longer hang around through an individual's lifetime, limiting his potential and keeping him on the lower margins of society. The already well-off may connect even more vigorously with their innate talent for empathy, and be more devoted to helping people on the bottom rise. Remembering the concept of social capital, introduced in chapter 5, please consider the possibility that neuroenablement may be a way in which we will all get rich.

This is why I've been working determinedly over the past several years to advance the cause of neurotechnology.

In late 2006 I formed the Neurotechnology Industry Organization (NIO), a trade association for neurotechnology. Prior to that I began hosting an annual conference for scientists, executives, and investors through my company, NeuroInsights. At every gathering of industry leaders, I share this idea with as many of them as possible—asking the U.S. government to directly push research in nonmilitary applications of neuroscience, specifically in developing treatments for brain and nervous system conditions ranging from dementia and depression to traumatic brain injury.[3] We also need funding devoted to the study of the societal implications, including ethical and legal questions, of advancing neurotechnology—especially around neuroenablement and neuroenhancement.

V. S. Ramachandran expressed the greatest reason for launching these efforts: "There's a lot of hype connected with some reports of neuroscience experiments, but also some very important and good information. The implications of neuroscience reach in the direction of everything under the sun, really. You think of it as the coming neurosociety, but it may be better to call it neurocosmos."

TEN

OUR EMERGING NEUROSOCIETY

Just as an undersea quake starts a mammoth tidal wave that builds until it can inundate a coastline, every time our ancestors invented new tools that vastly extended their control of the world around them, they eventually built a new epoch, a massive shift: New capabilities led to new industries, and then to deep transformations of social, economic, and political organizations, while also creating new modes of artistic and cultural expression.

We are now feeling the initial shock waves of another massive shift. The impact of neurotechnology is leading to enhancements of human life that will be at least as far-reaching in their effects as the changes brought about by the plow, steam engine, electricity, or space flight. The neurosociety is emerging, in our time. Like the gigantic shifts of humanity's past, our emerging neurosociety is a wild card. It holds enormous, seemingly equal promise for inducing an age of bliss or a living nightmare. The reason so much more is at stake in this epochal change is that our newest tools will give us nothing less than increasingly precise control over the most powerful factor in our lives—our own minds.

If you think about it for a minute, and look from a certain angle, everything humanity has ever done or invented has been aimed at gaining control of minds. Hunting strategies and technologies ease the fear of starvation. Distilled spirits lift our spirits, though short-term and with some gnarly risks. Religion, music, visual art, architectural design, athletic competitions, great cuisine, are all among the ways we've sought to make ourselves feel either better protected, or better connected, or more in tune with our best thinking and deepest potentials.

One day, not far from now, we may experience lasting freedom from many of the limitations imposed on us by our brain processes, which haven't changed much since the Paleolithic period. Then, we were hunter-gatherers. Fire was one of our peak technologies, along with our first "composite tool," patiently fashioned spear tips of rock lashed with rawhide to wooden shafts. Population density probably averaged about one person per square mile. But even though our world has gone through several revolutionary social, cultural, and technological changes since then, we're using very much the same organ of reasoning now, in a world of 6.6 billion long-living and highly interconnected people.

It was no picnic for our ancestors to scratch for roots and berries, and to hunt and be hunted by their gigantic and deadly co-inhabitants of the Paleolithic period, and to seek shelter in cold, damp caves. But twenty-first-century living involves an almost ceaseless barrage of excessive strain, fear, and overstimulation. It drives our minds relentlessly, makes them hurt like a perpetually clenched muscle that, metaphorically speaking, becomes stronger but also more prone to cramps and spasms. Our minds frequently become our own opponents, operating from within our defenses, generating new problems faster than they can find solutions to the old ones.

In our emerging neurosociety, which I expect to arrive in full over the next thirty years, you will eventually be able to continuously shape your emotional stability, sharpen your mental clarity, and extend your most desirable sensory states until they become your dominant experience of reality.

Cognitive liberty, brain privacy, the freedom to think and feel what you want without government or corporate intrusion—these will be the civil rights battles in our emerging neurosociety. People are already considering these questions, and framing the topics for public debate. They are called neuroethicists. Leaders in this field work at Harvard University, the University of Pennsylvania, and Stanford University to understand and clarify the emerging ethical issues. For example: Will governments have the right to subject criminal suspects to brain scans before they are proven guilty? Could a judge mandate a mind-altering "treatment" instead of a prison sentence? Will neurotechnology be used to control thoughts and actions that are generally deemed undesirable?

Here is what I believe to be the overarching question: Does a citizen's right to privacy include his or her inner domain of thought? Depending on how we answer questions like these, the emerging technologies may be used to control us and keep us in cultural or economic bondage. Or, instead, they may be used to enrich our lives through enhancements that tap into and expand some now-dormant positive potential we all have. It will be a dystopia, a utopia, or some blend of the two, kept in flux by the sense something better is still possible.

Like previous waves of societal change, the Neuro Revolution is being driven by the development of new low-cost technologies, specifically biochips that uncover the inner workings of cells, and brain imaging. The convergence of these two innovations is now making clear to us how the brain works—both on the inside, molecular level and on the systemwide, WHOLE-brain scale. We are already seeing a transformation in disease diagnosis and therapeutic development.

In recent years, the decreasing cost of biochips has made it possible to discover a large number of neurotransmitters, receptors, ion channels, and other proteins critical for normal brain function. At the same time, higher-resolution brain imaging technologies have made it easier to understand the what, when, and where of the electrical and chemical events that occur in our brains and form our thoughts and behaviors.

As the convergence of these technologies accelerates, the diverse and specific manifestations of neurotechnology will bloom, just as personal computing and the Internet gradually blossomed from the microchip.

The 1990s were dubbed "the Decade of the Brain," but thanks to recent neurotech developments, we have actually learned more about the brain in the past ten years than over the previous fifty.

Looking forward, we can see the neurotech marketplace will be expanded as the specificity and effectiveness of new drugs, devices, and diagnostics increase. But we can also be sure that the story will not end there. The same newly acquired knowledge about how to heal the sick or injured brain will make it possible to improve the performance of "normal" brains.

Given the elevated place of self-improvement in the human psyche, it seems inevitable the "lifestyle-improvement" market will drive an almost unlimited market expansion for neurotechnology. But as with all technologies, the sale and distribution of neurotechnology across humanity will be uneven. As a consequence, the forces of inequality and resentment that are already ripping at our hopes for peace will become even more powerful. Eventually these technologies will become fairly inexpensive. But that will take years, and until it happens the chasm between haves and have-nots will be obvious and dramatic. Large-scale violence will be a constant threat.

The vast unevenness and uneasiness of human existence became apparent to me in 1984, when I was a thirteen-year-old boy traveling home with my mother to Cupertino, California, from a six-week-long meditation retreat we had attended in Ganeshpuri, India. I was used to spending weekends playing computer games like Zork and Adventure on a home-based terminal linked to massive mainframe computers connected by the then fledgling Internet. I remember asking my dad, who managed the development of the ARPANET, what else computers would be used for. Among his list of research applications, most of which I can't remember, was a prediction that someday in the not-too-distant future this technology would usher in an era of "two-way TV."

Of course, the Internet has transformed a whole lot more in our lives over the past quarter century than just TV. But at the time I heard this prediction, this was a big enough change in my mind to stick with me in an important way.

As my mother and I rode through New Delhi on our way to catch our Air India flight, I peered out the window and saw glorious new buildings being constructed along unpaved streets. Spread across the construction sites and into the streets there were thousands of men, women, and children living in cardboard tents with their cows.

I felt very fortunate to have been born into the world I lived in.

But it wasn't until we landed in Dubai for a layover that it all came together. When we disembarked the 747 and entered an all-marble airport with Rolex clocks on the walls, numbered with embedded diamonds instead of black paint, I realized that the magnitude of the disparity that exists among humans would one day get uncovered with the help of "two-way TV," and it would cause unimaginable friction, eventually leading to wars. Needlessly to say, as someone in the first year of his teens, I wasn't able to articulate this idea specifically. But to this day I realize my glimpse of those diamond-encrusted wall clocks has shaped my life's journey so far.

In 2004, exactly twenty years later, and after I had spent a few years of researching and writing this book, the crown prince of Dubai, Sheikh Mohammed bin Rashid al-Maktoum, invited me back to Dubai to give the keynote dinner talk at the Arab Strategy Forum. Joining the fifty leading Arab businessmen, politicians, and technocrats were Bill Clinton, Madeleine Albright, Wesley Clark, and Thomas Friedman. We had all come together to discuss what the Arab world could look like in 2020, and how to help get it there. And it was with them that I first shared how our emerging neurosociety was going to reshape global culture in the decades to come.

With this diffusion of neurotechnology, a new form of human society will emerge, a postindustrial, postinformational neurosociety. The highlight of our neurosociety will be the tools that will help make living in our highly connected, urbanized world not only tolerable but also possibly magnificent. To explain, I'd like to introduce a term from psychology that is worth knowing about: "priming."

Priming is the idea that someone's attitudes or concepts can be activated by subtle cues, without that person's conscious awareness. For example, the political consultant Dr. Frank Luntz describes an impactful accident in his 2007 book, *Words That Work: It's Not What You Say, It's What People Hear*. Luntz gained fame for coming up with phrases that can shift the emotional tone of a hot-button political topic, like calling inheritance tax "the death tax," or casting drilling for oil as "energy exploration." These bits of verbal sleight-of-hand are one example of priming, though they're an obvious and fundamental, even primitive, technique. One day, Luntz discovered a far more subtle priming method.

In 1992 he was showing focus groups three short films of then-presidential candidate Ross Perot. The first was a biography; the second comprised some testimonials praising Ross Perot. The last one was a recorded speech by Perot himself. In one session, Luntz inadvertently showed the speech first. Afterward, he was stunned to find people in that group were far more negative about Perot than all of his previous groups. So he probed the phenomenon more deeply.

Further testing showed that viewing the speech first usually created a negative impression of the candidate. Luntz attributed this to the fact that Perot had an impressive business background, and was well respected for his successes, but didn't necessarily communicate this by his personal presence and his words. Instead, he communicated the fact that he was generally used to being obeyed. While his style was probably effective in corporate meetings, where he was resolutely in charge and didn't need to sell ideas so much as decree them, for a candidate who was seeking popular approval he seemed like an oddball, and an autocrat. His ideas were a bit different from those of typical politicians too. As Luntz puts it, "Unless and until you knew something about the man and his background, you would get the impression his mental tray was not quite in the full, upright, and locked position."

Thus, it wasn't the *information* received about Perot that mattered; it was the *order in which it came*. Put the speech last and he seemed to be a provocative spokesperson. Put it first and he looked like a toss-up between a dingbat and a full-blown wacko.

In one sense, this perceptual shift isn't too surprising. Sales and marketing are a process. You wouldn't expect a salesperson to attempt to close the deal before assessing the customer's needs, describing the product's benefits, and answering objections. In another sense, the subjects were passively viewing information of three different types. They all saw every bit of the content, and there was no interaction to "close" the deal. Nevertheless, the order of viewing made a huge difference in their opinions.

So the way we think about things can be highly malleable, and it can be reshaped even when we aren't aware of anyone's intention to give it a specific form. A stereotypically slick salesperson, or an obviously partisan political spokesperson using loaded phrases, will raise our defenses and crank up our BS filters to maximum levels. But something we scarcely notice may come along and completely alter our outlook.

Very recently, researchers found that something as subtle as a two-foot difference in ceiling height can alter the way the brain works. Lead researcher Joan Meyers-Levy recently explained, "When people are in a room with a high ceiling, they activate the idea of freedom. In a low-ceilinged room, they activate more constrained, confined concepts." Meyers-Levy is a professor of marketing at the University of Minnesota. She says that when we experience the concept of freedom, the information processing in our brains encourages a greater variation in the kinds of thoughts we will consider. If we feel confined, we unconsciously shift to more detail-oriented processing.

The study consisted of three tests ranging from anagram puzzles to product evaluation. In each test, a ten-foot ceiling correlated with subject activity that the researchers interpreted as "freer, more abstract thinking," whereas subjects in an eight-foot room were more likely to focus on specifics. Whenever we experience anxiety, we tend to hyperfocus, to inspect the smallest details and to draw overexcited conclusions.

This Minnesota study poses a whole cascade of intriguing possibilities. If a two-foot difference in ceiling height, a change that most people would not find noticeable enough to mention, can change how our brains operate, what could happen from manipulating other characteristics of our environments? How does a soaring cathedral ceiling affect our thoughts, compared to a flat ceiling? How does being inside a Frank Gehry structure composed of swooping curves affect our minds, compare to being inside a conventional rectilinear building? Does a windowless room engender different ideas than an office with a typical modest window, or one with floor-to-ceiling glass? What perceptual tendencies can we change by alterations in surrounding colors and textures?

The number of variables is huge, and few have been studied, but neuroimaging opens the possibility of figuring out these puzzles at low cost. Instead of building millions and millions of dollars' worth of test structures and painstakingly noting typical behaviors over a span of time, we now have the option of scanning an individual's brain while he or she views a vast array of rooms using virtual reality.

This means that every advance we gain in knowledge can more quickly ignite further advances. And so on, and so on.

Neuromarketing aside, retail marketers have long employed architectural priming techniques. Most of these have been fairly obvious, like building banks to make them look as substantial as possible—often as masonry buildings with classical pillars to connote timeless stability. Retailers of expensive clothing or other high-end merchandise create store environments with high-concept designs and high-quality flooring and fixtures so you can feel that their classiness, their prestige, will rub off on you when you buy their goods.

Most of this work has been done intuitively. We can expect that store concepts, workplaces, schools, and individual dwellings will soon be tested and tweaked using the tools of neuroscience. There are already people walking among us who define themselves as neuroarchitects.

Neuroarchitects, like the neurofinanciers, neuroestheticians, neuroeconomists, neurolaw experts, neurotheologians, neurowarfare researchers, and other thoroughly modern men and women you've encountered in these pages, are the advance troops in a march forward that the whole world is about to take. There are legitimately scary possibilities ahead, and we should begin right now the work of knowing how to manage and minimize them. But overall, I hope we can enter the emerging neurosociety with a sort of positive self-fulfilling prophecy, like the attitude I once heard Magic Johnson express at the end of a game, when an interviewer asked his thoughts about the next big challenge coming up: "I expect it to be *beautiful*."

ACKNOWLEDGMENTS

A book like this does not happen alone or overnight. It is the product of inspiration that emerges from the fascinating lives of the individuals conducting research across the globe and the people reporting on their daily progress. I spent nearly a decade poring over their work, and what an amazing journey it has been.

Throughout my travels, both inside my mind and across the world, many individuals have played meaningful roles at different points in the process. Many thanks go to Allen Scott, James Canton, Martin Greenberger, Michael Rothschild, Paul Zak, Jack Lindsay, Wrye Sententia, Garen Staglin, Ross Mayfield, Sam Barondes, Amy Cortese, Chris Lynch, Hylton Jolliffe, Curt Alexander, Noel Ekstrom, Josh McCarter, Matt Mahoney, Daniel Ritter, Frank Eeckman, and Paul Stimers. I owe special thanks to my agents Joe Spieler and Deirdre

Mullane for seeing my complete vision in the early script and my publisher Phil Revzin who drove me that last step deeper into the future.

I owe a special debt of gratitude to Byron Laursen who helped bring *The Neuro Revolution* to life. I brought Byron on board after several years of writing alone to help me make the lives of those I was interviewing sing more harmoniously. In the end I got much more. He is a magnificent storyteller, a great collaborator, and a truly wonderful person.

Most of all, I am eternally grateful for my friend, mentor, wife and love, Casey. She supported this book from its inception eight years ago and has never lost her precious enthusiasm and brilliant insight. She explored with me, challenged my thinking, and inspired a full vision of humanity's future. I truly look forward to living out this revolution with her and Kyle by my side.

San Francisco
December 2008

NOTES

Introduction: Into a Narrow Tunnel

1. Motoko Rich, "Oliver Sacks Joins Columbia Faculty as 'Artist,'" *New York Times*, 1 September 2007.

Chapter 2: The Witness on Your Shoulders

1. Adam Liptak, "U.S. Imprisons One in 100 Adults, Report Finds," *New York Times*, 29 February 2008.
2. Committee to Review the Scientific Evidence on the Polygraph, National Research Council, *The Polygraph and Lie Detection* (National Academies Press, 2003).
3. Strategic Intelligence, "Statement of the Director of Central Intelligence on the Clandestine Services and the Damage Caused by Aldrich Ames," Department of Political Science

at Loyola College in Maryland, http://www.loyola.edu/dept/politics/intel/dec95dci.html, accessed 7 August 2008.

4. Beth Orenstein, "Guilty? Investigating fMRI's Future as a Lie Detector," *Radiology Today* 6, no. 10 (16 May 2005):30.
5. Richard Willing, "MRI Tests Offer Glimpse at Brains Behind the Lies," *USA Today*, 26 June 2006.
6. Ronald Bailey, "Is Commercial Lie Detection Set to Go?" *Reason Online*, 27 February 2007, http://www.reason.com/news/show/118819.html, accessed 21 October 2008.

Chapter 3: Marketing to the Mind

1. Stuart Elliot, "Is the Ad a Success? The Brain Waves Tell All," *New York Times*, 31 March 2008.
2. Ali McConnon, "If I Only Had a Brain Scan," *BusinessWeek*, 16 January 2007.
3. Marco Iacoboni, Joshua Freedman, and Jonas Kaplan, "This Is Your Brain on Politics," *New York Times*, 11 November 2007.
4. Niknil Swaminathan, "This Is Your Brain on Shopping," *Scientific American*, http://www.sciam.com/article.cfm?id=this-is-your-brain-on-sho, accessed 7 August 2008.
5. Jonathan Leake and Elizabeth Gibney, "High Price Makes Wine Taste Better," [London] *Sunday Times*, 13 January 2008.

Chapter 4: Finance with Feelings

1. Jon Gertner, "The Futile Pursuit of Happiness," *New York Times*, 7 September 2003.
2. Adam Levy, "Sex, Drugs, Money: The Pleasure Principle," *International Herald Tribune*, 2 February 2006.

Chapter 6: Do You See What I Hear?

1. Michael J. Bannisy and Jamie Ward, "Mirror-Touch Synesthesia Is Linked with Empathy," *Nature Neuroscience* 10 (17 June 2007): 816.
2. V. S. Ramachandran and William Hirstein, "The Science of Art: A Neurological Theory of Aesthetic Experience," *Journal of Consciousness Studies* 6, no. 6–7 (June–July 1999).

Chapter 7: Where Is God?

1. Benedict Carey, "A Neuroscientific Look at Speaking in Tongues," *New York Times*, 7 November 2006.
2. TED, "Talks: Jill Bolte Taylor: My Stroke of Insight," Technology, Entertainment, Design, http://www.ted.com/index.php/talks/jill_bolte_taylor_s_powerful_stroke_of_insight.html, accessed 8 August 2008.
3. Asheim Hansen and E. Brodtkorb, "Partial Epilepsy with 'Ecstatic' Seizures," *Epilepsy Behavior* 4, no. 6 (December 2003): 667–73.
4. Sandra Blakeslee, "Out-of-Body Experience? Your Brain Is to Blame," *New York Times*, 3 October 2006.

Chapter 8: Fighting Neurowarfare

1. Denise Gellene and Karen Kaplan, "They're Bulking Up Mentally," *Los Angeles Times*, 20 December 2007.
2. James Randerson, "Scary or Sensational? A Machine That Can Look into the Mind," [London] *Guardian*, 6 March 2008.
3. Noah Shachtman, "Be More Than You Can Be," *Wired*, March 2003.

Chapter 9: Perception Shift

1. Anjan Chatterjee, "Cosmetic Neurology and Cosmetic Surgery: Parallels, Predictions, and Challenges," *Cambridge Quarterly of Healthcare Ethics* 16, 129. (April 2007)
2. Hedweb, "The Hedonistic Imperative," http://www.hedweb.com/index.html, accessed 8 August 2008.
3. Nikhil Swaminathan, "Legislation Introduced to Spur Treatments for Brain Ailments," *Scientific American*, 8 May 2008.

BIBLIOGRAPHY

Ackerman, Sandra J. *Hard Science, Hard Choices: Facts, Ethics, and Policies Guiding Brain Science Today.* New York: Dana Press, 2006.

Alper, Matthew. *The God Part of the Brain: A Scientific Interpretation of Human Spirituality and God.* New York: Rogue Press, 2006.

Alston, Brian. *What Is Neurotheology?* BookSurge Publishing, 2007.

Arthur, Brian. "Is the Information Revolution Dead?" *Business 2.0*, March 2002.

Bailey, Ronald. *Liberation Biology: The Scientific and Moral Case for the Biotech Revolution.* New York: Prometheus Books, 2005.

Bainbridge, William. "The Evolution of Semantic Systems in the Coevolution of Human

Potential and Converging Technologies." *Annals of the New York Academy of Sciences* 1013 (2004): 150–77.

Banissy, Michael J., and Jamie Ward. "Mirror-Touch Synesthesia Is Linked with Empathy." *Nature Neuroscience* 10 (17 June 2007): 815–16

Barnes, D. A. "CNS Drug Discovery: Realizing the Dream." *Drug Discovery World* (Summer 2002): 54–57.

Barondes, Samuel H. *Better Than Prozac: Creating the Next Generation of Psychiatric Drugs*. New York: Oxford University Press, 2001.

Bartels, Andreas, and Semir Zeki. "Functional Brain Mapping During Free Viewing of Natural Scenes." *Human Brain Mapping* 21, no. 2 (2003): 75–83.

Bell, Daniel. *The Coming of the Post-Industrial Society: A Venture into Social Forecasting*. New York: Basic Books, 1973.

Beniger, James. *The Control Revolution: Technological and Economic Origins of the Information Society*. Cambridge, MA: Harvard University Press, 1986.

Blakeslee, Sandra. "Brain Experts Now Follow the Money." *New York Times Magazine*, 7 September 2003.

Boire, Richard Glen. "On Cognitive Liberties Part III." *Journal of Cognitive Liberties* 2 (2000): 7–22.

Brand, Stewart. *The Media Lab: Inventing the Future at MIT*. New York: Penguin, 1988.

Braun, Stephen. *The Science of Happiness: Unlocking the Mysteries of Mood*. New York: Wiley, 2000.

Brizendine, Louann. *The Female Brain*. New York: Morgan Road Books, 2006.

Burn, Tom. "Student Perceptions of Methylphenidate Abuse at a Public Liberal Arts College." *Journal of American College Health* 49 (2000): 143–45.

Camerer, Colin, George Lowenstein, and D. Prelec. "Neuroeconomics: How Neuroscience Can Inform Economics." *Journal of Economic Literature* 43 (March 2005): 9–64.

Canham, L. T. "Silicon Technology and Pharmaceutics—An Impending Marriage in the Nanoworld." *Drug Discovery World* (Summer 2000): 56–63.

Caplan, Arthur. "Is Better Best?" *Scientific American* (September 2003): 68–73.

Carlsson, Arvid. "A Paradigm Shift in Brain Research." *Science* 294: 1021–24.

Castells, Manuel. *The Rise of the Network Society.* Oxford, UK: Blackwell Publishers, 1996.

Chatterjee, Anjan. "Cosmetic Neurology and Cosmetic Surgery: Parallels, Predictions, and Challenges." *Cambridge Quarterly of Healthcare Ethics* 16 (2007): 129–37

Chatterjee, Anjan. "Cosmetic Neurology: The Controversy over Enhancing Movement, Mentation, and Mood." *Neurology* 63 (2004): 968–74.

Chiu, P. H., T. M. Lohrenz, and P. R. Montague. "Smokers' Brains Compute, but Ignore, a Fictive Error Signal in a Sequential Investment Task. *Nature Neuroscience* 11, no. 4 (2008): 514–20.

Chorost, Michael. *Rebuilt: How Becoming Part Computer Made Me More Human.* New York: Houghton Mifflin, 2005.

Condon, Richard. *The Manchurian Candidate.* Four Walls Eight Windows, 2003.

Damasio, Antonio R. *Looking for Spinoza: Joy, Sorrow, and the Feeling Brain*, New York: Harvest Books, 2003.

D'Aquili, Eugene G., and Andrew B. Newberg. *The Mystical Mind: Probing the Biology of Religious Experience (Theology and the Sciences).* Augsburg Fortress Publishers, 1999.

Dennett, Daniel C. *Freedom Evolves.* New York: Viking, 2003.

Dowd, Kevin. "Too Big to Fail? Long-Term Capital Management and the Federal Reserve." *Cato Briefing Paper* no. 52 (1999).

Drexler, Eric K. *Engines of Creation: The Coming Era of Nanotechnology.* New York: Anchor, 1986.

"The Ethics of Brain Science: Open Your Mind." *Economist*, May 23, 2002, 77–79.

Ekman, Paul. *Emotions Revealed: Recognizing Faces and Feelings to Improve Communication and Emotional Life.* New York: Henry Holt, 2003.

Farah, Martha. "Neurocognitive Enhancement: What Can We Do and What Should We Do?" *Nature Reviews Neuroscience*, May 2004, 421–24.

Fehmi, Les, and Jim Robbins. *The Open-Focus Brain: Harnessing the Power of Attention to Heal Mind and Body.* New York: Trumpeter, 2007.

Fellous, Jean-Marc, and Michael A. Arbib, ed. *Who Needs Emotions? The Brain Meets the Robot.* New York: Oxford University Press, 2005.

Freeman, Christopher. *Long Waves in the World Economy.* London: Frances Pinter, 1983.

Freeman, Christopher, and Francisco Louçã. *As Time Goes By: From the Industrial Revolution to the Information Revolution*. Oxford, UK: Oxford University Press, 2001.

Freeman, Christopher, John Clark, and Luc Soete. *Unemployment and Technical Innovation: A Study of Long Waves and Economic Development*. London: Frances Pinter, 1982.

Freud, Sigmund. *Civilization and Its Discontents*. New York. Penguin, 2002.

Fukuyama, Francis. *Our Posthuman Future: The Consequences of the Biotechnology Revolution*. New York: Farrar, Straus and Giroux, 2002.

Garreau, Joel. *Radical Evolution: The Promise and Peril of Enhancing Our Minds, Our Bodies—and What It Means to Be Human*. New York: Doubleday, 2004.

Gazzaniga, Michael S. *The Ethical Brain*. New York: Dana Press, 2005.

———. *The Social Brain*. New York: Basic Books, 1985.

Gertner, Jon. "The Futile Pursuit of Happiness." *New York Times*, 2003.

Gibson, William. *Neuromancer: Remembering Tomorrow*. New York: Ace Books, 1984.

Gilbert, Daniel. *Stumbling on Happiness*. New York: Alfred A. Knopf, 2006.

Gilbert, Daniel T., and Tim D. Wilson. "Miswanting: Some Problems in the Forecasting of Future Affective States." In J. Forgas, ed., *Thinking and Feeling: The Role of Affect in Social Cognition*. Cambridge, UK: Cambridge University Press, 2000.

Glimcher, Paul W. "Decisions, Decisions, Decisions: Choosing a Biological Science of Choice." *Neuron* 36, no. 2 (October 2002): 323–32.

———. *Decisions, Uncertainty, and the Brain: The Science of Neuroeconomics*. Cambridge, MA: MIT Press/Bradford Press, 2002.

Greenfield, Susan. *Tomorrow's People: How 21st Century Technology Is Changing the Way We Think and Feel*. London: Allen Lane, 2003.

Harvey, David. *Conditions of Postmodernity*. Cambridge, UK: Blackwell, 1989.

Huxley, Aldous. *Brave New World*. Garden City, NY: Doubleday, 1932.

———. *The Doors of Perception*. New York: Harper and Row, 1970.

———. *Island*. New York: Harper Perennial, 2002.

Illes, Judy, Allyson C. Rosen, Lynn Huang, R. A. Goldstein, Thomas A. Raffin, G. Swan, and Scott W. Atlas. "Ethical Consideration of Incidental Findings on Adult Brain MRI in Research." *Neurology* 62, no. 6 (2004): 888–90.

James, William. *The Varieties of Religious Experience: A Study in Human Nature.* 1902. New York: Routledge, 2002.

Kane, Pat. *The Play Ethic: A Manifesto for a Different Way of Living.* London: Pan Books, 2006.

Kant, Immanuel. *The Critique of Judgment.* New York: Cosimo Classics, 2007.

Kauffman, Stuart. *The Origins of Order: Self-Organization and Selection in Evolution.* New York: Oxford University Press, 1993.

Kelly, Kevin. *Out of Control: The Rise of Neo-Biological Civilization.* New York: Addison-Wesley, 1994.

Kesey, Ken. *One Flew Over the Cuckoo's Nest.* New York: Signet, 1963.

Key, Wilson B. *Subliminal Seduction.* New York: Signet, 1973.

Knutson, Brian, Charles S. Adams, Grace W. Fong, and Daniel Hommer. "Anticipation of Monetary Reward Selectively Recruits Nucleus Accumbens." *Journal of Neuroscience* 21 (2001): RC159.

Knutson, Brian, Jamil Bhanji, Rebecca E. Cooney, Lauren Atlas, and Lan H. Gotlib. "Neural Responses to Monetary Incentives in Major Depression." *Biological Psychiatry* 63 (2008): 686–92.

Knutson, Brian, Jeffrey Burgdorf, and Jaak Panksepp. "High-Frequency Ultrasonic Vocalizations Index Conditioned Pharmacological Reward in Rats." *Physiology and Behavior* 66 (1999): 639–43.

Knutson, Brian, and Peter Bossaerts. "Neural Antecedents of Financial Decisions." *Journal of Neuroscience* 27 (2007): 8174–77.

Kolman, Joe. "LTCM Speaks." *Derivatives Strategy Magazine,* April 1999, 25–32.

Kondratieff, Nikolai D. *The Long Wave Cycle.* New York: Richardson and Snyder, 1984.

Kuhn, Thomas. *The Structure of Scientific Revolutions.* 2nd ed. Chicago: The University of Chicago Press, 1970.

Kuhnen, Camelia, and Brian Knutson. "The Neural Basis of Financial Risk Taking." *Neuron* 47 (2005): 763–70.

Kurzweil, Ray. *The Singularity Is Near: When Humans Transcend Biology.* New York: Viking, 2005.

Kuznets, Simon. *Economic Change.* New York: W.W. Norton, 1953.

Langleben, Daniel, Frank M. Dattilio, and Thomas G. Guthei. "True Lies: Delusions and Lie-Detection Technology." *Journal of Psychiatry and Law* 34, no. 3 (2006): 351–70.

LeDoux, Joseph. *The Emotional Brain: The Mysterious Underpinnings of Emotional Life.* New York: Touchstone, 1996.

Lehrer, Jonah. *Proust Was a Neuroscientist.* New York: Houghton Mifflin, 2007.

Levitin, Daniel J. *This Is Your Brain on Music: The Science of a Human Obsession.* New York: Plume, 2007.

Levy, Adam. "Brain Scans Show Link Between Lust for Sex and Money." Bloomberg.com, February 1, 2006.

Lo, Andrew W., and Dmitri V. Repin. "The Psychophysiology of Real-Time Financial Risk Processing." *Journal of Cognitive Neuroscience* 14 (2002): 323–39.

Loewenstein, George. "Emotions in Economic Theory and Economic Behavior." *American Economic Review: Papers and Proceedings* 90 (2000): 426–32.

Loewenstein, George, and Daniel Adler. "A Bias in the Prediction of Tastes." *Economic Journal* 105 (1995): 929–37.

Loewenstein, George, Ted O'Donoghue, and Matthew Rabin. "Projection Bias in Predicting Future Utility." *Quarterly Journal of Economics* 118 (2003): 1209–48.

Luntz, Frank. *Words That Work: It's Not What You Say, It's What People Hear.* New York: Hyperion, 2006.

Lynch, Casey, and Zack Lynch. *The Neurotechnology Industry 2008: Drugs, Devices, and Diagnostics for the Brain and Nervous System; Market Analysis and Strategic Investment Guide to the Global Neurological Disease and Psychiatric Illness Markets.* San Francisco: NeuroInsights, 2008.

Lynch, Zack. "Emotions in Art and the Brain." *Lancet Neurology* 3 (2004): 191.

———. "Neuropolicy (2005–2035): Converging Technologies Enable Neurotechnology, Creating New Ethical Dilemmas." In William S. Bainbridge and Mihail C. Roco, eds. *Managing Nano-Bio-Info-Cogno Innovations: Converging Technologies in Society.* The Netherlands: Springer, 173–91.

———. "Neurotechnology and Society 2010–2060 in the Coevolution of Human Potential and Converging Technologies." *Annals of the New York Academy of Sciences* 1013 (2004): 229–33.

Malone, Thomas W. *The Future of Work: How the New Order of Business Will Shape Your Organization, Your Management Style, and Your Life*. Boston: HBS Press, 2004.

McCabe, Kevin, and Vernon Smith. "A Two Person Trust Game Played by Naïve and Sophisticated Subjects." *Proceedings of the National Academy of Sciences* 97, no. 7 (2000): 3777–81.

McClure, Samuel M., and P. Read Montague. "Neural Correlates of Behavioral Preference for Culturally Familiar Drinks." *Neuron* 44, no. 2 (October 14, 2004): 379–87.

McKinney, Laurence O. *Neurotheology: Virtual Religion in the 21st Century*. American Institute for Mindfulness, 1994.

McLuhan, Marshall. *Understanding Media—The Extensions of Man*. Cambridge, MA: MIT Press, 1964.

Mithen, Steven. *The Singing Neanderthals: The Origins of Music, Language, Mind, and Body*. Cambridge, MA: Harvard University Press, 2005.

Montague, P. Read. "Neuroeconomics: A View from Neuroscience." *Functional Neurology* 22, no. 4 (2007): 219–34.

Montague, P. Read, and Pearl Chiu. "For Goodness' Sake." *Nature Neuroscience* 10, no. 2 (2007): 137–38.

Montague, Read. *Why Choose This Book? How We Make Decisions*. New York: Penguin, 2006.

Moreno, Jonathan D. *Mind Wars: Brain Research and National Defense*. New York: Dana Press, 2006.

———. *Undue Risk: Secret State Experiments on Humans*. New York: Routledge, 2000.

Nabokov, Vladimir. *Speak, Memory*. New York: Everyman's Library, 1999.

Negropante, Nicholas P. *Being Digital*. New York: Vintage Books, 1995.

Newberg, Andrew. *Why We Believe What We Believe: Uncovering Our Biological Need for Meaning, Spirituality, and Truth*. New York: Free Press, 2006.

Newberg, Andrew, Eugene D'Aquili, and Vince Rause. *Why God Won't Go Away: Brain Science and the Biology of Belief*. New York: Ballentine Books, 2002.

Newberg, Andrew, et al. "The Measurement of Regional Cerebral Blood Flow During the Complex Cognitive Task of Meditation: A Preliminary SPECT Study." *Psychiatry Research* 106, no. 2 (2001): 113–22.

Oeppen, Jim W. "Broken Limits to Life Expectancy." *Science* 296 (2002): 1029–31.

Packard, Vance. *The Hidden Persuaders*. New York: D. Mackay, 1957.

Panskeep, Jaak. *Affective Neuroscience: The Foundations of Human and Animal Emotions*. New York: Oxford University Press, 1998.

Parens, Erik, ed. *Enhancing Human Traits: Ethical and Social Implications*. Washington, DC: Georgetown University Press, 1998.

Penrose. Roger. *The Emperor's New Mind: Concerning Computers, Minds, and the Laws of Physics*. New York: Oxford University Press, 1989.

Perez, Carlota. *Technological Revolutions and Financial Capital: The Dynamics of Bubbles and Golden Ages*. Northampton, MA: Edward Elgar, 2002.

Peterson, Richard L. *Inside the Investor's Brain: The Power of Mind over Money*. New York: Wiley Trading, 2007.

Phelps, Elizabeth A., et al. "Neurophysiological Mechanisms Underlying the Understanding and Imitation of Action." *Nature Reviews Neuroscience* 12, no. 5 (2000): 729–38.

Pinker, Steven. *The Blank Slate: The Modern Denial of Human Nature*. New York. Penguin Press, 2002.

Porter, Michael E. *The Competitive Advantage of Nations*. New York: Free Press, 1990.

Postrel, Virginia. *The Substance of Style: How the Rise of Aesthetic Value Is Remaking Commerce, Culture, and Consciousness*. New York: HarperCollins, 2003.

President's Council on Bioethics. *Beyond Therapy: Biotechnology and the Pursuit of Happiness*. Washington, DC: President's Council on Bioethics, 2003.

Ramachandran, V. S. "Mirror Neurons and Imitation Learning as the Driving Force Behind the Great Leap Forward in Human Evolution." Third Edge. www.edge.org.

Ramachandran, V. S., and Sandra Blakeslee. *Phantoms in the Brain: Human Nature and the Architecture of the Mind*. New York: Fourth Estate, 1998.

Rich, Motoko. "Oliver Sacks Joins Columbia Faculty as 'Artist.'" *New York Times*, 1 September 2007.

Roco, Mihail. "Science and Technology Integration for Increased Human Potential and Societal Outcomes in the Coevolution of Human Potential and Converging Technologies." *Annals of the New York Academy of Sciences* 1013 (2004): 1–16.

Roco, Mihail C., and William S. Bainbridge, eds. *Converging Technologies for Improving Human Performance: Nanotechnology, Biotechnology, Information Technology, and Cognitive Science*. 2002.

Rose, Steven. *The Future of the Brain: The Promise and Perils of Tomorrow's Neuroscience.* Oxford, UK: Oxford University Press, 2005.

Rothschild, Michael. *Bionomics: Economy as Ecosystem.* New York: Henry Holt, 1990.

Sacks, Oliver. *Awakenings.* New York: Duckworth, 1973.

———. *Musicophilia: Tales of Music and the Brain.* New York: Alfred A. Knopf, 2007.

Safire, William. "Neuroethics: Mapping the Field, a Report from the Conference." *Cerebrum* 4, no. 3 (Summer 2002).

Sandel, Michael J. "The Case Against Perfection: What's Wrong with Designer Children, Bionic Athletes, and Genetic Engineering." *Atlantic Monthly*, April 2004, 58.

Schopenhauer, Arthur. *The World as Will and Representation.* New York: Longman, 2007.

Schumpeter, Joseph. *Business Cycles: A Theoretical, Historical, and Statistical Analysis of the Capitalist Process.* 2 vols. New York and London: McGraw-Hill, 1939.

Schwartz, Peter. *The Art of the Long View.* New York: Doubleday, 1991.

Scripture, Edward W. *The New Psychology.* 1898. Kessinger Publishing, 2007.

Sententia, Wrye. "Brain Fingerprinting: Databodies to Databrains." *Journal of Cognitive Liberty* 2, no. 3 (2001): 31–46.

———. "Neuroethical Considerations: Cognitive Liberty and Converging Technologies for Improving Human Cognition." In The Coevolution of Human Potential and Converging Technologies. *Annals of the New York Academy of Sciences* 1013 (2004): 221–28.

Sheridan, Clare. "Benefits of Biotech Clusters Questioned." *Nature Biotechnology* 21, no. 11 (2003): 1258–59.

Shermer, Michael. *The Mind of the Market: Compassionate Apes, Competitive Humans, and Other Tales from Evolutionary Economics.* New York. Henry Holt, 2008.

Smith, Adam. *An Inquiry into the Nature and Causes of the Wealth of Nations.* 1776. New York: Bantam, 2004.

Stock, Gregory. *Metaman: The Merging of Humans and Machines into a Global Superorganism.* New York: Simon and Schuster, 1993.

———. *Redesigning Humans: Our Inevitable Genetic Future.* New York: Houghton Mifflin, 2002.

Tarnas, Richard. *Cosmos and Psyche: Intimations of a New World View*. New York: Plume, 2006.

Taylor, Jill B. *My Stroke of Insight: A Brain Scientist's Personal Journey*. New York: Viking, 2008.

Tinbergen, Nikolaas. *The Herring Gull's World: A Study of the Social Behaviors of Birds*. London: Anchor Books, 1967.

Toffler, Alvin. *Future Shock*. New York: Bantam Books, 1970.

———. *The Third Wave*. New York: Bantam Books, 1980.

Wolpe, Paul R., Kenneth R. Foster, and Daniel D. Langleben. "Emerging Neurotechnologies for Lie Detection: Promises and Perils." *American Journal of Bioethics* 5, no. 2 (2005): 39–49.

Zak, Paul. "The Neurobiology of Trust." In *Proceedings of the 2003 Economic Science Association Conference*.

Zeki, Semir. "Artistic Creativity and the Brain." *Science* 293(2001): 51–52.

———. "The Chronoarchitecture of the Human Brain." *NeuroImage* 22, no. 1 (2004): 419–33.

———. *Inner Vision: An Exploration of Art and the Brain*. New York: Oxford University Press, 1999.

Zimmer, Carl. *Soul Made Flesh: The Discovery of the Brain—and How It Changed the World*. New York: Free Press, 2004.

INDEX